Selenium Essentials

Get to grips with automated web testing with
the amazing power of Selenium WebDriver

Prashanth Sams

BIRMINGHAM - MUMBAI

Selenium Essentials

Copyright © 2015 Packt Publishing

All rights reserved. No part of this book may be reproduced, stored in a retrieval system, or transmitted in any form or by any means, without the prior written permission of the publisher, except in the case of brief quotations embedded in critical articles or reviews.

Every effort has been made in the preparation of this book to ensure the accuracy of the information presented. However, the information contained in this book is sold without warranty, either express or implied. Neither the author, nor Packt Publishing, and its dealers and distributors will be held liable for any damages caused or alleged to be caused directly or indirectly by this book.

Packt Publishing has endeavored to provide trademark information about all of the companies and products mentioned in this book by the appropriate use of capitals. However, Packt Publishing cannot guarantee the accuracy of this information.

First published: March 2015

Production reference: 1230315

Published by Packt Publishing Ltd.
Livery Place
35 Livery Street
Birmingham B3 2PB, UK.

ISBN 978-1-78439-433-2

www.packtpub.com

Credits

Author
Prashanth Sams

Reviewer
Alen Šiljak

Commissioning Editor
Pramila Balan

Acquisition Editor
Llewellyn Rozario

Content Development Editor
Sweny Sukumaran

Technical Editor
Parag Topre

Copy Editor
Sarang Chari

Project Coordinator
Rashi Khivansara

Proofreaders
Safis Editing
Maria Gould

Indexer
Hemangini Bari

Production Coordinator
Melwyn D'sa

Cover Work
Melwyn D'sa

About the Author

Prashanth Sams is a test automation engineer contributing to the IT industry since 2011. He graduated with a bachelor's degree in information technology from Anna University and lives in Chennai, India, with his family. He started his career as a human resource executive. Later, he worked at an HR outsourcing (US recruiting) company that operates in Chennai.

He is very passionate about test automation and has chosen to be a professional software engineer. He is an active blogger and a moderator for `http://seleniumworks.blogspot.in/`, a blog about Selenium, and is a great supporter of the Selenium Community, responding diligently to questions and answers over professional networks. He loves emerging technologies with soft skills development and spends 14 to 16 hours a day on them. In a short span of time, he has gained rich experience in various projects, handling different automation tools. Prashanth's Twitter handle is `@prashanthsams`.

I would like to thank all Selenium Core committers and the Selenium Community members who spend most of their time making this open source product a successful tool. I would also like to thank the editors of this book, who are very intense and responsible for bringing about knowledge transfer to software professionals.

About the Reviewer

Alen Šiljak is a solutions architect and software development enthusiast who was lucky enough to live through the times from the 8-bit machines to the 64-bit ones. As a fan of Agile methodologies, he appreciates creativity and enjoys creating order from entropy. Currently, he is happy to see software become mainstream, but he still sighs for the times when technologies were obsolete in a matter of months, if not weeks. Also, he still marvels at the fact that his phone is incomparably more advanced than the machines on which he started his IT journey.

Thank you, my family and friends. You are the icing on the cake of my life.

www.PacktPub.com

Support files, eBooks, discount offers, and more

For support files and downloads related to your book, please visit www.PacktPub.com.

Did you know that Packt offers eBook versions of every book published, with PDF and ePub files available? You can upgrade to the eBook version at www.PacktPub.com and as a print book customer, you are entitled to a discount on the eBook copy. Get in touch with us at service@packtpub.com for more details.

At www.PacktPub.com, you can also read a collection of free technical articles, sign up for a range of free newsletters and receive exclusive discounts and offers on Packt books and eBooks.

https://www2.packtpub.com/books/subscription/packtlib

Do you need instant solutions to your IT questions? PacktLib is Packt's online digital book library. Here, you can search, access, and read Packt's entire library of books.

Why subscribe?

- Fully searchable across every book published by Packt
- Copy and paste, print, and bookmark content
- On demand and accessible via a web browser

Free access for Packt account holders

If you have an account with Packt at www.PacktPub.com, you can use this to access PacktLib today and view 9 entirely free books. Simply use your login credentials for immediate access.

Table of Contents

Preface

Selenium WebDriver is an open source software-testing tool used to automate web-based applications that is platform independent and that can be accessed by any popular programming languages. It's been about a decade since Jason Huggins started the Selenium project in 2004 at Thoughtworks. Later, in 2008, Simon Stewart combined his work on WebDriver with Selenium to give a new birth to Selenium WebDriver. Today, Selenium WebDriver is the most widely used web-automation tool around the world.

This book provides guidance that will help readers grasp Selenium WebDriver concepts fast. You will learn about the advanced features of the Selenium IDE and Selenium Builder, followed by cross-browser tests, methods of Selenium WebDriver, best practices involved, and extensive ideas to create a Selenium framework.

What this book covers

Chapter 1, *The Selenium IDE*, provides intense ideas to practice record-and-playback IDEs such as the Selenium IDE and Selenium Builder.

Chapter 2, *Selenium WebDriver Cross-browser Tests*, helps you to do efficient compatibility tests. Here, we will also learn about how to run tests in the cloud.

Chapter 3, *Selenium WebDriver Functions*, delivers all the functions of Selenium WebDriver in detail with examples on each.

Chapter 4, *Selenium WebDriver Best Practices*, explains how to manage Selenium automation tasks with dissimilar techniques.

Chapter 5, *Selenium WebDriver Frameworks*, guides you to customize and build any kind of automation framework using Selenium WebDriver.

What you need for this book

- Microsoft Windows
- MAC / Ubuntu (Linux)
- Eclipse IDE/IntelliJ IDEA
- Selenium IDE
- Selenium Builder
- Mozilla Firefox
- Google Chrome
- Internet Explorer
- Opera
- Apple Safari
- Microsoft Excel

Who this book is for

Selenium Essentials is intended for software professionals who want to learn about Selenium WebDriver from scratch and for testers who want to migrate from Selenium RC to Selenium WebDriver. This book delivers an easy learning curve for Selenium newbies who want to begin with Selenium WebDriver and a perfect guide for intermediate Selenium testers to become masters in Selenium WebDriver.

Conventions

In this book, you will find a number of styles of text that distinguish between different kinds of information. Here are some examples of these styles, and an explanation of their meaning.

Code words in text, database table names, folder names, filenames, file extensions, pathnames, dummy URLs, user input, and Twitter handles are shown as follows: "Now, replace the directory on your code with `C:/Users/user_name/AppData/Local/Google/Chrome/New User`."

A block of code is set as follows:

```
<testdata>
  <testvarname="value" />
  <testvarname="value" />
  <testvarname="value" />
</testdata>
```

When we wish to draw your attention to a particular part of a code block, the relevant lines or items are set in bold:

```
driver.switchTo().frame(value);
```

Any command-line input or output is written as follows:

```
$ Unzip chromedriver_linux64.zip
$ cp chromedriver /usr/local/bin
$ chmod +x /usr/local/bin/chromedriver
```

New terms and **important words** are shown in bold. Words that you see on the screen, in menus or dialog boxes for example, appear in the text like this: "It will prompt you to enter the **admin** password; enter it to set the path."

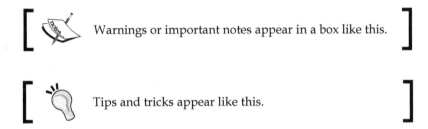

Warnings or important notes appear in a box like this.

Tips and tricks appear like this.

Reader feedback

Feedback from our readers is always welcome. Let us know what you think about this book—what you liked or may have disliked. Reader feedback is important for us to develop titles that you really get the most out of.

To send us general feedback, simply send an e-mail to feedback@packtpub.com, and mention the book title via the subject of your message.

If there is a topic that you have expertise in and you are interested in either writing or contributing to a book, see our author guide on www.packtpub.com/authors.

Customer support

Now that you are the proud owner of a Packt book, we have a number of things to help you to get the most from your purchase.

Downloading the example code

You can download the example code files for all Packt books you have purchased from your account at http://www.packtpub.com. If you purchased this book elsewhere, you can visit http://www.packtpub.com/support and register to have the files e-mailed directly to you.

Errata

Although we have taken every care to ensure the accuracy of our content, mistakes do happen. If you find a mistake in one of our books—maybe a mistake in the text or the code—we would be grateful if you would report this to us. By doing so, you can save other readers from frustration and help us improve subsequent versions of this book. If you find any errata, please report them by visiting http://www.packtpub.com/submit-errata, selecting your book, clicking on the **errata submission form** link, and entering the details of your errata. Once your errata are verified, your submission will be accepted and the errata will be uploaded on our website, or added to any list of existing errata, under the Errata section of that title. Any existing errata can be viewed by selecting your title from http://www.packtpub.com/support.

Piracy

Piracy of copyright material on the Internet is an ongoing problem across all media. At Packt, we take the protection of our copyright and licenses very seriously. If you come across any illegal copies of our works, in any form, on the Internet, please provide us with the location address or website name immediately so that we can pursue a remedy.

Please contact us at copyright@packtpub.comwith a link to the suspected pirated material.

We appreciate your help in protecting our authors, and our ability to bring you valuable content.

Questions

You can contact us at questions@packtpub.com if you are having a problem with any aspect of the book, and we will do our best to address it.

1
The Selenium IDE

The **Selenium IDE (Integrated Development Environment)** is an open source record-and-playback tool for generating Selenium scripts, which is integrated with the Firefox web browser as an extension. It is a renowned web-based UI test automation tool that extracts any kind of locator from the web page. The locators can be either attribute-based or structure-based, and include ID, name, link, XPath, CSS, and DOM. The IDE has the entire Selenium Core, which allows the users to record, playback, edit, and debug tests manually in a browser. The user actions in the web page can be recorded and exported in any of the most popular languages, such as Java, C#, Ruby, and Python.

Selenium Builder is an alternative open source tool for the Selenium IDE to record and playback web applications. It is an extension of the Firefox web browser, which is similar to the Selenium IDE, but, it has some unique features that the Selenium IDE doesn't support. Selenium Builder is a standard tool from Sauce Labs that runs tests on Sauce Cloud from the Selenium Builder interface itself.

In this chapter, we will learn about:

- Selenium IDE's record and playback abilities
- Selenium IDE functions
- Selenium IDE Data Driven tests
- Selenium IDE JavaScript functions
- Selenium Builder record and playback
- Selenium Builder Data Driven tests
- Selenium Builder on cloud

The Selenium IDE is a Firefox extension to record and playback web-based applications. However, it does more than what a record-and-playback tool would do. Breakpoints allow the users to debug IDE commands step by step on runtime. The IDE has three different types of panes, namely the left pane, test case pane, and log / reference / UI-element / rollup pane.

Launch the Selenium IDE from the Firefox **Tools** menu, **Tools | Selenium IDE**. The IDE can also be opened using the *Ctrl + Shift + S* shortcut or by clicking on the Selenium icon in the top-right corner of the Firefox web browser. The Selenium icon is shown in the following screenshot:

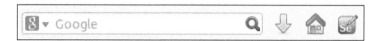

A new, untitled test case will be created in **Left Pane** after launching the Selenium IDE. To start with a new test case, choose **New Test Case** from the **File** menu, that is, **File | New Test Case**, or make use of *Ctrl + N*, the Windows shortcut.

To start recording test scripts, click on the round, red icon from the playback control toolbar. By default, the record button will be active and the test scripts are recorded in **Selenese**, a domain-specific language that is similar to the HTML format. The playback control toolbar is shown in the following screenshot:

The **Fast-Slow** slider adjusts the test speed execution; the **Play All** button lets you run entire test cases as a test suite, where a test suite is a collection of test cases; and the **Play** button helps you to run the current test case. The **Pause/Resume** button pauses test execution for a while and allows the user to resume tests at their convenience.

The **Test Case** pane displays all the recorded steps with **Command**, **Target**, and **Value**. The **Command** column instructs the IDE about what to do. It comes with three different aspects, which are:

- Actions
- Accessors
- Assertions

The Selenium IDE has a list of built-in commands that let you drive tests as expected. Adding user-defined commands to the Selenium IDE is quite feasible by extending the external JavaScript methods. A command can be any one of the preceding three types. In the IDE, these commands are easily editable and replaceable with alternative commands while generating scripts.

Action commands manipulate the application state through some kind of actions, can be either `action` or `actionAndWait`. Action commands that end with the suffix `AndWait` allow the page to load fully before starting to execute the next command.

A few of the Action command examples are `open`, `type`, `typeAndWait`, `select`, `selectAndWait`, `check`, `checkAndWait`, `click`, and `clickAndWait`.

Accessors detect the application state and store results in a variable; `store`, `storeText`, and `storeValue` are the commands that are used to store values. For example, in the following screenshot, `search` is a variable and `prashanth sams` is the search keyword. Later on, the stored value is retrieved and is used as a parameter for an action, `${search}`. The discussion in this paragraph is encapsulated in this screenshot:

Command	Target	Value
open	/?gfe_rd=cr&ei=...	
store	prashanth sams	search
type	css=#gbqfq	${search}

Assertions verify the application state by validating the expected result. It is available in three different modes, namely, **assert**, **verify**, and **waitFor**. Assert fails and aborts the test execution upon failure, verify fails and continues the test execution upon failure, and waitFor waits for a specific condition to occur and fails upon timeout. By default, the timeout is set to 30,000 milliseconds, 30 seconds. In the Selenium IDE, the timeout can be manually configured using the **Options** menu.

The **Target** field directs the IDE to locate elements, and the general syntax for **Target** is as follows:

```
locatorType = argument
```

An example for **Target** is as follows:

```
css=#gbqfq
```

The **Log** / **Reference** / **UI-Element** / **Rollup** pane is displayed at the bottom of the IDE. This is shown in the following screenshot:

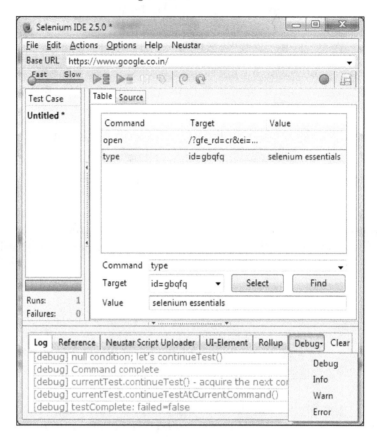

This pane allows the user to see the log information, command reference, **UI-Element**, and **Rollup**, among others. On installing the Neustar plugin to the Selenium IDE, a separate tab called **Neustar Script Uploader** will be shown along with the other tabs. **Neustar WPM** (formerly called **Browsermob**) is a web performance management tool for web page monitoring and load testing.

The log captures all the IDE test execution steps one by one and is mainly used for debugging purposes. The **Debug** menu in the bottom pane contains a list of options, namely **Info**, **Debug**, **Warn**, and **Error**. It lets you filter the explicit status, warning, and error messages, and certainly reduces the verbosity level.

The **Reference** tab gives a detailed explanation of the IDE commands upon clicking on each row from the **Test Case** pane. In the case of user-defined commands, the **Reference** tab will not include any information. **Rollup** executes a group of commands in one step; it is reusable and can be used any number of times within the test case. Refer to **Help | UI-Element Documentation** for more details about **UI-Element** and **Rollup**.

While recording test scripts, the Selenium IDE provides UI-based options for every mouse right-click on elements on a browser web page. To achieve this, right-click on the web page and hover the mouse over **Show All Available Commands**. The following screenshot is the result of this action:

WebDriver playback

The WebDriver playback feature in Selenium IDE lets you run tests in any one of the most popular web browsers: Chrome, Firefox, HtmlUnit, Internet Explorer, and Opera. By default, the WebDriver playback feature is turned off and is inactive. To run Selenium IDE scripts through WebDriver, turn on the WebDriver playback settings.

Launch the Selenium IDE and choose **Selenium IDE Options** from the **Options** menu. Switch to the **WebDriver** tab and select the **Enable WebDriver** checkbox. Now, restart the Selenium IDE to enable the WebDriver playback feature. However, on changing the browser name, restarting the IDE is not necessary. The idea discussed in these two paragraphs is shown in the following screenshot:

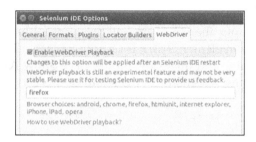

Prerequisites for the WebDriver playback feature

The following are the prerequisites that need to be fulfilled to enable the WebDriver playback feature:

- Download the latest Selenium Server standalone library (JAR) file
- Install Java to start the Selenium Server
- Download the latest drivers for popular browsers (**chromedriver**, **IEDriver**, and so on)

Selenium Server can be initialized manually from the terminal or Command Prompt. Open the terminal or Command Prompt, locate the Selenium Server JAR file, and run the command using the following syntax:

```
java -jar selenium-server-standalone-<version-number>.jar
```

Now you can run the command:

```
java -jar selenium-server-standalone-2.44.0.jar
```

Click on the Selenium IDE **Play** ▶= button to drive tests through WebDriver. To run tests in the Chrome web browser, replace the text `firefox` with `chrome` under the Selenium IDE options, as shown in the following screenshot:

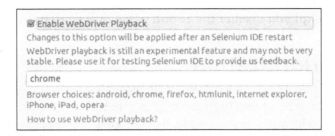

By default, it is essential to set the **ChromeDriver** path in your working machine. Download the latest ChromeDriver extension from `http://chromedriver.storage.googleapis.com/index.html?path=2.9/`.

Follow these configuration steps to set the ChromeDriver extension path for different platforms.

On Windows:

1. Double-click and open the **My Computer** window.

2. Right-click anywhere on the window and select **Properties**.

3. Click on **Advanced System settings**.

4. Click on **Environment Variables** from **System Properties**.

5. Under **System variables**, select the variable named **Path** and click on the **Edit** button.

6. Now, extract the downloaded ChromeDriver package and copy the location path.

7. Paste the extracted location in **Path** (under **System variables**) and click on **OK**.

On Linux:

Open the terminal and run the following command:

```
$ wget http://chromedriver.storage.googleapis.com/2.7/chromedriver_
linux64.zip
$ Unzip chromedriver_linux64.zip
$ cp chromedriver /usr/local/bin
$ chmod +x /usr/local/bin/chromedriver
```

On Mac:

1. Unzip/extract the zipped package (`chromedriver_mac32.zip`).
2. Copy and paste ChromeDriver to `/usr/bin`.
3. It will prompt you to enter the **admin** password; enter it to set the path.

Locator prioritization

Prioritization lets you prioritize locators while recording scripts. In general, this feature helps the user by giving high priority to generate scripts with respect to the user's preferred locators. For example, by changing the `csslocator` order from the fifth to the first position, the further elements generated on the Selenium IDE's target will be in CSS, that is, the `locatorType` will be set to CSS by default.

An example of this is `CSS = argument`.

Launch the Selenium IDE, choose **Options...** from the **Options** menu, and switch to the **Locator Builders** tab. The left-hand pane will be mounted with a list of available locator builders, such as, `ui`, `id`, `link`, `name`, `css`, `dom:name`, `xpath:link`, `xpath:img`, `xpath:attributes`, `xpath:idRelative`, `xpath:href`, `dom:index`, and `xpath:position`. The list of available locator builders is shown in the following screenshot:

Drag and drop locator builders on the left-hand side to change their order. Finally, click on the **OK** button, and restart the Selenium IDE for the changes to take effect. To reset the default settings of the Selenium IDE, click on the **Reset** option found in the bottom-left corner of the **Selenium IDE options** window pane.

Avoiding Selenium export

Exporting test cases each and every time can bother the user. The Selenium IDE provides an excellent feature to avoid such exporting trouble by using a quick solution. In general, clicking on the **Source** toggle button under the Test Case pane displays the current test case in the Selenese language. The Selenium IDE transforms the existing Selenese language into a user-preferred script format, such as, **Java/JUnit4/WebDriver**.

Launch the Selenium IDE and choose **Options...** from the **Options** menu. Make sure that the option **Enable experimental features** is selected and click on the **OK** button. Click on the **Format** option in the **Options** menu and select the preferred combination format.

For example, you can select the **Java/JUnit4/WebDriver** option, as shown in the following screenshot.

Finally, restart the Selenium IDE for the changes to take effect. In the Selenium IDE, there is no support for exporting the test cases in TestNG with the WebDriver (**Java/TestNG/WebDriver**) combination format.

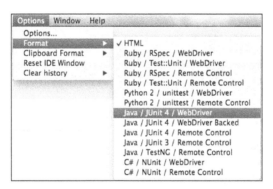

The tab will automatically switch to the **Source** view on disabling the **Table** toggle button, as shown in the following screenshot:

```
import java.util.regex.Pattern;
import java.util.concurrent.TimeUnit;
import org.junit.*;
import static org.junit.Assert.*;
import static org.hamcrest.CoreMatchers.*;
import org.openqa.selenium.*;
import org.openqa.selenium.firefox.FirefoxDriv
import org.openqa.selenium.support.ui.Select;

public class Untitled {
  private WebDriver driver;
  private String baseUrl;
  private boolean acceptNextAlert = true;
  private StringBuffer verificationErrors = ne

  @Before
```

The Selenium IDE clipboard

Copying snippets through **Clipboard Format** is one of the quickest ways to obtain instantly generated scripts. Here, a snippet may contain one or two lines of code. The following screenshot displays different types of export formats available under the **Clipboard Format** option:

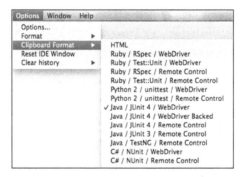

Launch the Selenium IDE, hover your mouse over the **Clipboard Format** option from the **Options** menu, and select the preferred combination format.

An example of a combination format is **Java/JUnit 4/WebDriver**. In general, the HTML snippet is set as default. The following screenshot shows a row being copied from the **Test Case** pane:

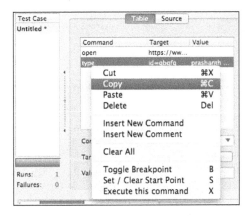

Copy the following row from the **Test Case** pane (as shown in the preceding screenshot) and paste it as a code snippet:

```
driver.findElement(By.id("gbqfq")).clear();
driver.findElement(By.id("gbqfq")).sendKeys("prashanthsams");
```

Data Driven tests

Parameterization is a part of the Data Driven technique for retrieving values from an external data source as input. In general, the Data Driven tests are used to verify the actual and expected values from an external data source. The Selenium IDE plays a major role in parameterization, as it operates with different sets of permutations and combinations. Let's see how we can use a JavaScript file as a data source for Data Driven tests. The following is the JavaScript syntax for parameterization:

```
varname = "value"
```

For example, create a JavaScript (.js) file (Datasource.js) that includes the following keywords:

```
Search1 = "PrashanthSams"
Search2 = "Selenium Essentials"
```

Launch the Selenium IDE and choose **Options...** from the **Option** menu, **Option | Options...**. Now, browse through Selenium IDE extensions and select the .js file created earlier (Datasource.js), as shown in the following screenshot:

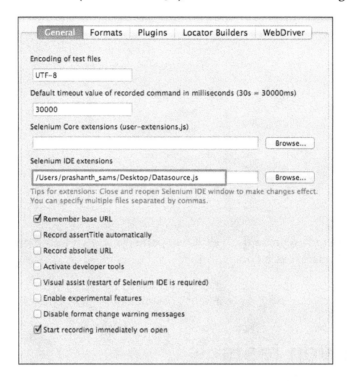

Finally, restart the Selenium IDE to effect the changes. Initialize, store, and fetch values from the .js file one by one using the storeEval command, as follows:

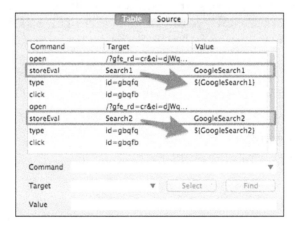

Here, Search1 and Search2 are the two variables that retrieve the respective keywords from the JavaScript file. These values are again stored in the new variables, GoogleSearch1 and GoogleSearch2, as shown in the preceding screenshot.

User-defined JavaScript methods

The IDE actions, accessors, and assertions can be user-defined and customized. To achieve this, the user is supposed to add JavaScript methods to the Selenium object prototype and the PageBot object prototype. In general, the Selenium IDE verifies the user-defined JavaScript methods on launching the IDE. Selenium Core extensions (user-extensions.js) in **Options...** give support to upload the user-defined JavaScript files.

Let's discuss this with an example that involves step-by-step instructions, as follows:

1. Refer to the Google site https://sites.google.com/site/ seleniumworks/selenium-ide-data-driven to download the following JavaScript files:

    ° datadriven.js

    ° goto_sel_ide.js

    ° user-extensions.js

2. Launch the Selenium IDE and choose **Options...** from the **Option** menu, **Option | Options....** Now, navigate to **Selenium Core extensions** and upload the JavaScript files (user-extensions.js, goto_sel_ide.js and datadriven.js), as shown in the following screenshot. Finally, restart the Selenium IDE for the changes to take effect.

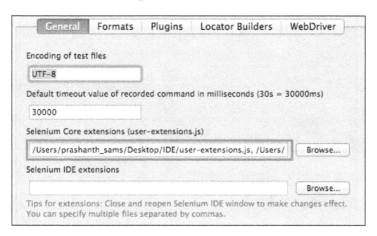

3. In the Selenium IDE, the XML file is used as a data source to store values, whereas `Datadriven.js` is designed to support the XML file format.

Here is the syntax for XML file formatted data source:

```
<testdata>
  <test varname="value" />
  <test varname="value" />
  <test varname="value" />
</testdata>
```

Create an XML file with the extension `.xml` (`data.xml`). Here, `varname` is the variable name, and `value` refers to the keyword under the `<test>` tag. Let's create an XML file with the `.xml` extension (`data.xml`). For example, refer to the following code snippet:

```
<testdata>
  <test phrase="selenium essentials" />
  <test phrase="seleniumworks.com" />
  <test phrase="prashanthsams" />
</testdata>
```

4. Take a look at these details: `loadTestData` is a user-defined command that fetches the XML data source, `while` and `endWhile` do looping, whereas `nextTestData` checks for the data from the next row in the data source. The user can add any number of JavaScript methods. The following screenshot shows this step in detail:

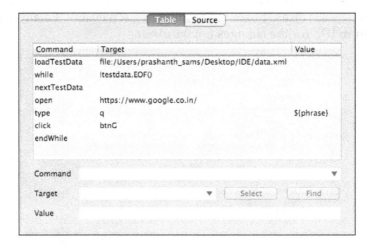

Selenium IDE JavaScript functions

In addition to user-defined JavaScript commands introduced through
`user-extensions.js`, the Selenium IDE allows the user to create JavaScript
queries or functions directly in the **Target** field. For example, let's run a Google
search by getting a random number between 1 and 100, as follows:

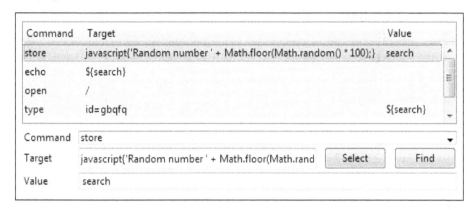

The following HTML source tags let you convert the steps into runnable test scripts:

```html
<tr>
  <td>store</td>
  <td>javascript{'Random number ' + Math.floor(Math.random() * 100);}
</td>
  <td>search</td>
</tr>
<tr>
  <td>echo</td>
  <td>${search}</td>
  <td></td>
</tr>
<tr>
  <td>open</td>
  <td>/</td>
  <td></td>
</tr>
<tr>
  <td>type</td>
  <td>id=gbqfq</td>
  <td>${search}</td>
</tr>
```

Simple JavaScript execution

The predefined Selenium IDE JavaScript command `runScript` is a very powerful command that lets the user execute simple JavaScript functions directly from the IDE, for example, `javascript{alert("Hello!")}`.

Let's see how we can disable an active textbox and enable an inactive textbox using the following code snippet:

```
document.getElementsByName('****')[0].setAttribute('disabled', '')
document.getElementsByName('****')[0].removeAttribute('disabled');
```

Command	Target	V...
open	https://accounts.google.com/	
runScript	document.getElementsByName('Passwd')[0].setAttribute('disabled', '')	
runScript	document.getElementsByName('Passwd')[0].removeAttribute('disabled');	

Mouse scroll

The `scroll` event is currently unavailable in the Selenium IDE. However, the `user-extensions.js` file includes a JavaScript method that lets you scroll the mouse through the web page.

Refer to the Google site `https://sites.google.com/site/seleniumworks/selenium-ide-tricks` to download `user-extensions.js`. This user extension file includes IDE commands like `while`, `endWhile`, `gotoIf`, `gotoLabel`, and `push`. Increase the value to 10 based upon the vertical length of the web page, as shown in the following screenshot:

Command	Target	Value
open	https://accounts.google.com/	
store	20	i
store	0	looptimes
while	storedVars.looptimes <= 10	
storeEval	selenium.browserbot.getCurrentWindow().scrollTo(0,${i})	
store	javascript{storedVars.looptimes++;}	
storeEval	${i}+20	i
endWhile		

Parameterization using arrays

The Selenium IDE command `storeEval` is used to store values in a variable while running scripts, whereas `storedVars` is a JavaScript associate array with string indexes containing variables. In the following example, `storeEval` reserves the list of rivers in an array, and `getEval` is a command for initiating and incrementing the values. Some of the commands used in this section are purely user-defined, such as `while`, `endWhile`, and so on. Here, the `endWhile` command is used to break the loop once the value inside the array reaches the maximum limit. The following screenshot gives us a clear idea of what is being discussed here:

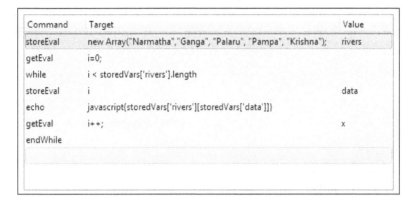

Command	Target	Value
storeEval	new Array("Narmatha","Ganga", "Palaru", "Pampa", "Krishna");	rivers
getEval	i=0;	
while	i < storedVars['rivers'].length	
storeEval	i	data
echo	javascript{storedVars['rivers'][storedVars['data']]}	
getEval	i++;	x
endWhile		

Let's see another example of advanced parameterization concepts using the Selenium IDE. Refer to the Google site `https://sites.google.com/site/seleniumworks/selenium-ide-tricks` to download the `user-extensions.js` file. The following screenshot captures the essence of this discussion:

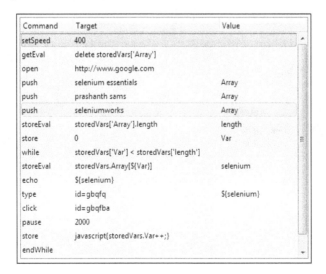

Command	Target	Value
setSpeed	400	
getEval	delete storedVars['Array']	
open	http://www.google.com	
push	selenium essentials	Array
push	prashanth sams	Array
push	seleniumworks	Array
storeEval	storedVars['Array'].length	length
store	0	Var
while	storedVars['Var'] < storedVars['length']	
storeEval	storedVars.Array[${Var}]	selenium
echo	${selenium}	
type	id=gbqfq	${selenium}
click	id=gbqfba	
pause	2000	
store	javascript{storedVars.Var++;}	
endWhile		

In this example, the values are pushed into the array manually, and it does a Google search one by one.

Selenium Builder

Selenium Builder is a record-and-playback tool similar to the Selenium IDE and is an extension of the Firefox web browser. It has some unique features that the Selenium IDE doesn't support, for example GitHub integration to export and commit test suites/cases in GitHub and TestingBot integration to run tests in the cloud. It also provides more language support than the Selenium IDE, including for languages such as JSON, Java/TestNG, NodeJS WD, NodeJS Mocha, and NodeJS Protractor. Selenium Builder is expected to be the future of the Selenium IDE, with advanced features. Here's a screenshot of Selenium Builder:

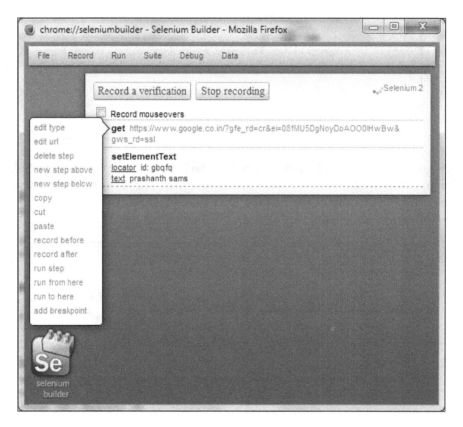

Recording and playback

Upon confirmation of the addition of the Selenium Builder extension into the Firefox browser, open the web page that is being tested (such as www.google.com). There are a number of ways to open Selenium Builder, such as:

- Right-click on the web page and select **Launch Selenium Builder**.
- Choose **Selenium Builder** from the **Tools** menu, that is, **Tools | Web Developer | Launch Selenium Builder**
- Alternatively, you can use the *Ctrl + Alt + B* shortcut keys

Selenium Builder is supposed to have an option **Selenium 2** to record WebDriver test scripts. The user can easily identify the page being actively recorded, as it has a green-colored tab, as shown in the following screenshot:

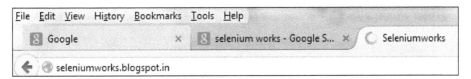

Selenium Builder allows you to have control over the web app to start recording test scripts. It allows the users to verify their tests using a button, **Record a verification**. Clicking on the verification button highlights the text to be verified all over the page on mouseover. The mouseover function is an excellent feature of Selenium Builder that helps the user to add mouse hover functionality whenever needed. Select the **Record mouseovers** checkbox to record mouseover actions. The following screenshot shows this functionality:

Stop recording the scripts and export (**File | Export**) the tests in the preferred combination format.

Data Driven tests

In general, Selenium Builder is capable of importing test scripts saved by the Selenium IDE, since both the IDEs' native format is Selenese. Exporting Selenium scripts in the **Java/TestNG/WebDriver** format, which avoids all the extra human effort when working with TestNG, is feasible in Selenium Builder. TestNG is one of the most popular unit testing frameworks and is similar to JUnit for Java bindings.

Despite the fact that Selenium Builder is a record-and-playback tool, it allows users to perform basic Data Driven tests by receiving input from the data source. The values can also be stored temporarily in Selenium Builder using the **Manual Entry** option (**Data | Manual Entry**). To do so, create a variable and assign the value, as shown in the following screenshot:

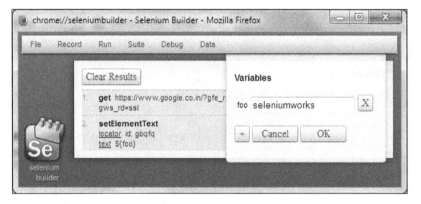

Selenium Builder comprises higher-level support for Data Driven testing that lets you drive tests using the JSON, XML, and CSV file formats. CSV file formats can be used to fetch data files with large volumes. Let's see some of the file formats involving Data Driven tests.

Testing using a JSON file

The following is the syntax for a JSON file formatted data source:

```
[
{ "varname": "value", "varname": "value", ... },
{ "varname": "value", "varname": "value", ... },
   ...
]
```

For example, create a JSON file with the `.json` extension (`data.xml`). Here, `foo` is the variable name, whereas `value` is defined with the keywords `prashanth` and `sams`, as follows:

```
[ {"foo": "prashanth"}, {"foo": "sams"}]
```

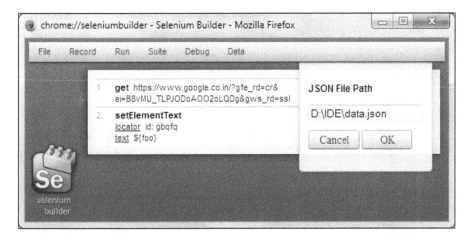

To address the values stored in a JSON file, it is recommended to call the variables, for example, `${foo}`, in the pre-recorded steps, as shown in the preceding screenshot.

Testing using an XML file

The following is the syntax for an XML file formatted data source:

```
<testdata>
  <testvarname="value" />
  <testvarname="value" />
  <testvarname="value" />
</testdata>
```

For example, create an XML file with the `.xml` extension (`data.xml`). Here, `search` is the variable name, and `value` refers to the keyword under the **<test>** tag, as shown in the following code snippet:

```
<testdata>
  <test search="selenium essentials" />
  <test search="prashanthsams" />
```

```
        <test search="seleniumworks" />
    </testdata>
```

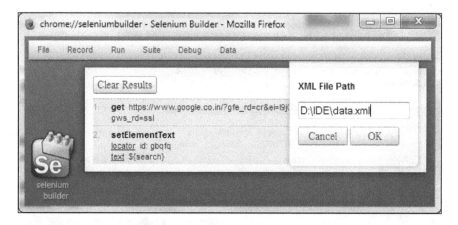

Selenium Builder on the cloud

Selenium Builder allows users to run cross-browser tests on the cloud directly from the IDE interface. To integrate Selenium Builder with Sauce and enable export and playback scripts on Sauce OnDemand, it's necessary to install the Selenium Builder Sauce plugin first. Launch Selenium Builder, click on **Manage plugins**, and install the Sauce for Selenium Builder plugin. Also, users should create a Sauce account before running cloud-based Selenium tests. For further details on Selenium Builder on the cloud, refer to `https://saucelabs.com/`. The following screenshot shows us the **Plugins** page:

Obtain the Sauce access key after logging in to your Sauce account as a real user or by clicking **look up access key** in Selenium Builder. The access key is unique for each user.

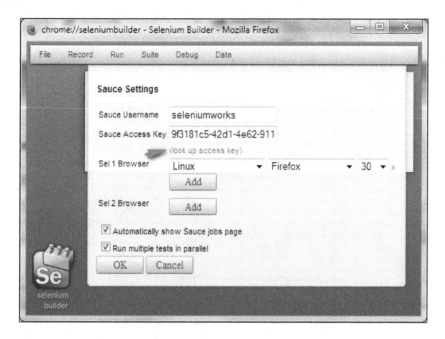

Choose **Run on Sauce OnDemand** from the **Run** menu. Make sure that **Sauce Settings** is furnished before running the test cases. All you need to do is enter the **Sauce Username**, **Sauce Access Key**, **Browser**, and **OS** versions. Finally, log in to Sauce and verify the test results. The tests are recorded in video and images for user preview.

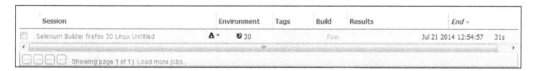

Summary

In this chapter, you learned about the Selenium IDE functions, along with Selenium Builder, and observed how to handle the Selenium IDE to automate simple tests.

In the next chapter, we will discuss advanced compatibility-testing techniques using Selenium WebDriver. It lets you drive tests on different browsers.

2
Selenium WebDriver Cross-browser Tests

The term cross-browser testing can be applied to both multi-browser testing and compatibility testing. Testing the web application with multiple web browsers is defined as cross-browser testing. A lack of cross-browser testing results in layout and functionality issues. Manually testing an application needs a lot of human effort and time to finish a complex job, but automated tests are carried out to avoid such issues. In general, most of the cross-browser issues are generated while rendering web page elements, which results in a functional and UI-based mess. Selenium WebDriver provides excellent support for automating test cases with the most popular browsers, using their own drivers. Selenium cross-browser tests can also be automated on the cloud using web application tools, such as SauceLabs, BrowserStack, and TestingBot.

In this chapter, we will cover the following topics:

- Selenium WebDriver compatibility tests
- Selenium cross-browser tests on cloud
- Selenium headless browser testing
- Switching user agents
- Tests on specific Firefox versions
- Tests from custom Firefox profile
- Tests from custom Chrome profile

Selenium WebDriver compatibility tests

Selenium WebDriver handles browser compatibility tests on almost every popular browser, including Chrome, Firefox, Internet Explorer, Safari, and Opera. In general, every browser's JavaScript engine differs from the others, and each browser interprets the HTML tags differently. The WebDriver API drives the web browser as the real user would drive it. By default, FirefoxDriver comes with the `selenium-server-standalone.jar` library added, but for Chrome, IE, and Opera, there are libraries that need to be added or instantiated externally.

Let's see how we can instantiate each of the following browsers through its own driver:

- **Mozilla Firefox**: The `selenium-server-standalone` library is bundled with FirefoxDriver to initialize and run tests in a Firefox browser. FirefoxDriver is added to the Firefox profile as a file extension on starting a new instance of FirefoxDriver. Please check the Firefox versions and its suitable drivers at `http://selenium.googlecode.com/git/java/CHANGELOG`.

  The following is the code snippet to kick start Mozilla Firefox:

  ```
  WebDriver driver = new FirefoxDriver();
  ```

- **Google Chrome**: Unlike FirefoxDriver, the ChromeDriver is an external library file that makes use of WebDriver's wire protocol to run Selenium tests in a Google Chrome web browser. The following is the code snippet to kick start Google Chrome:

  ```
  System.setProperty("webdriver.chrome.driver","C:\\chromedriver.exe");
  WebDriver driver = new ChromeDriver();
  ```

> To download ChromeDriver, refer to `http://chromedriver.storage.googleapis.com/index.html`.

- **Internet Explorer**: `IEDriverServer` is an executable file that uses the WebDriver wire protocol to control the IE browser in Windows. Currently, IEDriverServer supports the IE versions 6, 7, 8, 9, and 10. The following code snippet helps you to instantiate IEDriverServer:

  ```
  System.setProperty("webdriver.ie.driver","C:\\IEDriverServer.exe");
  ```

```
DesiredCapabilities dc = DesiredCapabilities.internetExplorer();
dc.setCapability(InternetExplorerDriver.INTRODUCE_FLAKINESS_BY_
IGNORING_SECURITY_DOMAINS, true);
WebDriver driver = new InternetExplorerDriver(dc);
```

 To download IEDriverServer, refer to `http://selenium-release.storage.googleapis.com/index.html`.

- **Apple Safari**: Similar to FirefoxDriver, SafariDriver is internally bound with the latest Selenium servers, and starts the Apple Safari browser without any external library. SafariDriver supports Safari browser version 5.1.x and runs only on Mac. For more details, refer to `http://elementalselenium.com/tips/69-safari`.

  The following code snippet helps you to instantiate SafariDriver:

  ```
  WebDriver driver = new SafariDriver();
  ```

- **Opera**: OperaPrestoDriver (formerly called OperaDriver) is available only for Presto-based Opera browsers. Currently, it does not support Opera versions 12.x and above. However, the recent releases (Opera 15.x and above) of Blink-based Opera browsers are handled using OperaChromiumDriver. For more details, refer to `https://github.com/operasoftware/operachromiumdriver`.

  The following code snippet helps you to instantiate OperaChromiumDriver:

  ```
  DesiredCapabilities capabilities = new DesiredCapabilities();
  capabilities.setCapability("opera.binary", "C://Program Files
  (x86)//Opera//opera.exe");
  capabilities.setCapability("opera.log.level", "CONFIG");
  WebDriver driver = new OperaDriver(capabilities);
  ```

 To download OperaChromiumDriver, refer to `https://github.com/operasoftware/operachromiumdriver/releases`.

TestNG

TestNG (**Next Generation**) is one of the most widely used unit-testing frameworks implemented for Java. It runs Selenium-based browser compatibility tests with the most popular browsers. The Eclipse IDE users must ensure that the TestNG plugin is integrated with the IDE manually. However, the TestNG plugin is bundled with IntelliJ IDEA as default. The testng.xml file is a TestNG build file to control test execution; the XML file can run through Maven tests using POM.xml with the help of the following code snippet:

```
<plugin>
  <groupId>org.apache.maven.plugins</groupId>
  <artifactId>maven-surefire-plugin</artifactId>
  <version>2.12.2</version>
  <configuration>
    <suiteXmlFiles>
      <suiteXmlFile>testng.xml</suiteXmlFile>
    </suiteXmlFiles>
  </configuration>
</plugin>
```

To create a testng.xml file, right-click on the project folder in the Eclipse IDE, navigate to **TestNG | Convert to TestNG**, and click on **Convert to TestNG**, as shown in the following screenshot:

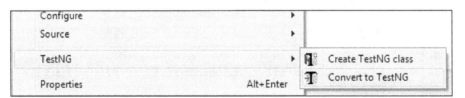

The testng.xml file manages the entire test. It acts as a mini data source by passing the parameters directly into the test methods. The location of the testng.xml file is shown in the following screenshot:

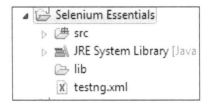

As an example, create a Selenium project (for example, `Selenium Essentials`) along with the `testng.xml` file, as shown in the previous screenshot. Modify the `testng.xml` file with the following tags:

```xml
<?xml version="1.0" encoding="UTF-8" ?>
<!DOCTYPE suite SYSTEM "http://testng.org/testng-1.0.dtd">
<suite name="Suite" verbose="3" parallel="tests" thread-count="5">

  <test name="Test on Firefox">
    <parameter name="browser" value="Firefox" />
      <classes>
        <class name="package.classname" />
      </classes>
  </test>

  <test name="Test on Chrome">
    <parameter name="browser" value="Chrome" />
    <classes>
      <class name="package.classname" />
    </classes>
  </test>

  <test name="Test on InternetExplorer">
    <parameter name="browser" value="InternetExplorer" />
    <classes>
      <class name="package.classname" />
    </classes>
  </test>

  <test name="Test on Safari">
    <parameter name="browser" value="Safari" />
    <classes>
      <class name="package.classname" />
    </classes>
  </test>

  <test name="Test on Opera">
    <parameter name="browser" value="Opera" />
    <classes>
      <class name="package.classname" />
    </classes>
  </test>
</suite>
<!-- Suite -->
```

Download all the external drivers except FirefoxDriver and SafariDriver, extract the zipped folders, and locate the external drivers as mentioned in the preceding snippets for each browser.

 Refer to *Chapter 1, The Selenium IDE,* to learn how to set the path on different platforms.

The following Java snippet will explain how you can get parameters directly from the testng.xml file and how you can run cross-browser tests as a whole:

```java
@BeforeTest
@Parameters({"browser"})
public void setUp(String browser) throws MalformedURLException {
  if (browser.equalsIgnoreCase("Firefox")) {
    System.out.println("Running Firefox");
    driver = new FirefoxDriver();
  } else if (browser.equalsIgnoreCase("chrome")) {
    System.out.println("Running Chrome");
    System.setProperty("webdriver.chrome.driver", "C:\\chromedriver.
exe");
    driver = new ChromeDriver();
  } else if (browser.equalsIgnoreCase("InternetExplorer")) {
    System.out.println("Running Internet Explorer");
    System.setProperty("webdriver.ie.driver", "C:\\IEDriverServer.
exe");
    DesiredCapabilities dc = DesiredCapabilities.internetExplorer();
    dc.setCapability
    (InternetExplorerDriver.INTRODUCE_FLAKINESS_BY_IGNORING_SECURITY_
DOMAINS, true);
    //If IE fail to work, please remove this line and remove enable
protected mode for all the 4 zones from Internet options
    driver = new InternetExplorerDriver(dc);
  } else if (browser.equalsIgnoreCase("safari")) {
    System.out.println("Running Safari");
    driver = new SafariDriver();
  } else if (browser.equalsIgnoreCase("opera")) {
    System.out.println("Running Opera");
  // driver = new OperaDriver();      --Use this if the location is
set properly--
    DesiredCapabilities capabilities = new DesiredCapabilities();
    capabilities.setCapability("opera.binary", "C://Program Files
    (x86)//Opera//opera.exe");
    capabilities.setCapability("opera.log.level", "CONFIG");
    driver = new OperaDriver(capabilities);
  }
}
```

SafariDriver is not yet stable. A few of the major issues in SafariDriver are as follows:

- SafariDriver won't work properly in Windows
- SafariDriver does not support modal dialog box interaction
- You cannot navigate forwards or backwards in the browser history through SafariDriver

Selenium cross-browser tests on the cloud

The ability to automate Selenium tests on the cloud is quite interesting, with instant access to real devices. Sauce Labs, BrowserStack, and TestingBot are the leading web-based tools used for cross-browser compatibility checking. These tools contain unique test automation features, such as diagnosing failures through screenshots and video, executing parallel tests, running Appium mobile automation tests, executing tests on internal local servers, and so on.

SauceLabs

SauceLabs is the standard Selenium test automation web app to do cross-browser compatibility tests on the cloud. It lets you automate tests in your favorite programming languages, using test frameworks such as JUnit, TestNG, Rspec, and many more. SauceLabs cloud tests can also be executed from the Selenium Builder IDE interface. Check for the available SauceLabs devices, OS, and platforms at `https://saucelabs.com/platforms`.

Access the website from your web browser, log in, and obtain the Sauce username and **Access Key**. Make use of the credentials to drive tests over the SauceLabs cloud. SauceLabs creates a new instance of the virtual machine while launching the tests. Parallel automation tests are also possible using SauceLabs. The following is a Java program to run tests over the SauceLabs cloud:

```
package packagename;

import java.net.URL;
import org.openqa.selenium.remote.DesiredCapabilities;
import org.openqa.selenium.remote.RemoteWebDriver;
import java.lang.reflect.*;

public class saucelabs {
```

```
private WebDriver driver;

@Parameters({"username", "key", "browser", "browserVersion"})
@BeforeMethod
public void setUp(@Optional("yourusername") String username,
                  @Optional("youraccesskey") String key,
                  @Optional("iphone") String browser,
                  @Optional("5.0") String browserVersion,
                  Method method) throws Exception {

    // Choose the browser, version, and platform to test
    DesiredCapabilities capabilities = new DesiredCapabilities();
    capabilities.setBrowserName(browser);
    capabilities.setCapability("version", browserVersion);
    capabilities.setCapability("platform", Platform.MAC);
    capabilities.setCapability("name", method.getName());
    // Create the connection to SauceLabs to run the tests
    this.driver = new RemoteWebDriver(
    new URL("http://" + username + ":" + key + "@ondemand.saucelabs.
    com:80/wd/hub"), capabilities);
}

@Test
public void Selenium_Essentials() throws Exception {
    // Make the browser get the page and check its title
    driver.get("http://www.google.com");
    System.out.println("Page title is: " + driver.getTitle());
    Assert.assertEquals("Google", driver.getTitle());
    WebElement element = driver.findElement(By.name("q"));
    element.sendKeys("Selenium Essentials");
    element.submit();
}
@AfterMethod
public void tearDown() throws Exception {
    driver.quit();
}
}
```

SauceLabs has a setup similar to BrowserStack on test execution and generates detailed logs. The breakpoints feature allows the user to manually take control over the virtual machine and pause tests, which helps the user to investigate and debug problems. By capturing JavaScript's console log, the JS errors and network requests are displayed for quick diagnosis while running tests against the Google Chrome browser.

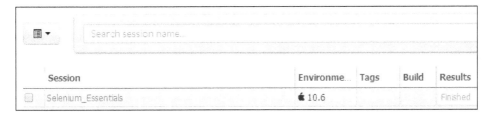

BrowserStack

BrowserStack is a cloud-testing web app to access virtual machines instantly. It allows users to perform multi-browser testing of their applications on different platforms. It provides a setup similar to SauceLabs for cloud-based automation using Selenium.

Access the site `https://www.browserstack.com` from your web browser, log in, and obtain the BrowserStack username and **Access Key**. Make use of the obtained credentials to drive tests over the BrowserStack cloud.

For example, the following generic Java program with TestNG provides a detailed overview of the process that runs on the BrowserStack cloud. Customize the browser name, version, platform, and so on, using capabilities. Let's see the Java program we just talked about:

```java
package packagename;

import org.openqa.selenium.remote.DesiredCapabilities;
import org.openqa.selenium.remote.RemoteWebDriver;

public class browserstack {

  public static final String USERNAME = "yourusername";
  public static final String ACCESS_KEY = "youraccesskey";
  public static final String URL = "http://" + USERNAME + ":" +
  ACCESS_KEY + "@hub.browserstack.com/wd/hub";

  private WebDriver driver;

  @BeforeClass
  public void setUp() throws Exception {
    DesiredCapabilities caps = new DesiredCapabilities();
    caps.setCapability("browser", "Firefox");
    caps.setCapability("browser_version", "23.0");
    caps.setCapability("os", "Windows");
```

```
      caps.setCapability("os_version", "XP");
      caps.setCapability("browserstack.debug", "true");
      //This enable Visual Logs

      driver = new RemoteWebDriver(new URL(URL), caps);
   }

   @Test
   public void testOnCloud() throws Exception {
      driver.get("http://www.google.com");
      System.out.println("Page title is: " + driver.getTitle());
      Assert.assertEquals("Google", driver.getTitle());
      WebElement element = driver.findElement(By.name("q"));
      element.sendKeys("seleniumworks");
      element.submit();
   }

   @AfterClass
   public void tearDown() throws Exception {
      driver.quit();
   }
}
```

The app generates and stores test logs for the user to access anytime. The generated logs provide a detailed analysis with step-by-step explanations. To enhance the test speed, run parallel Selenium tests on the BrowserStack cloud, but the automation plan has to be upgraded to increase the number of parallel test runs.

TestingBot

TestingBot also provides a setup similar to BrowserStack and SauceLabs for cloud-based cross-browser test automation using Selenium. It records a video of the running tests to analyze problems and debug them. Additionally, it provides support to capture the screenshots on test failure. To run local Selenium tests, it provides an SSH tunnel tool that lets you run tests against local servers or other web servers. TestingBot uses Amazon's cloud infrastructure to run Selenium scripts in various browsers.

Access the site https://testingbot.com/, log in, and obtain the **Client Key** and **Client Secret** from your TestingBot account. Make use of the credentials to drive tests over the TestingBot cloud.

Let's see an example Java test program with TestNG, using the Eclipse IDE that runs on the TestingBot cloud:

```java
package packagename;

import java.net.URL;

import org.openqa.selenium.remote.DesiredCapabilities;
import org.openqa.selenium.remote.RemoteWebDriver;

public class testingbot {
  private WebDriver driver;

  @BeforeClass
  public void setUp() throws Exception {
    DesiredCapabilitiescapabillities = DesiredCapabilities.firefox();
    capabillities.setCapability("version", "24");
    capabillities.setCapability("platform", Platform.WINDOWS);
    capabillities.setCapability("name", "testOnCloud");
    capabillities.setCapability("screenshot", true);
    capabillities.setCapability("screenrecorder", true);
    driver = new RemoteWebDriver(
    new URL
    ("http://ClientKey:ClientSecret@hub.testingbot.com:4444/wd/hub"),
      capabillities);
  }

  @Test
  public void testOnCloud() throws Exception {
    driver.get
      ("http://www.google.co.in/");
    driver.findElement(By.id("q")).clear();
    WebElement element = driver.findElement(By.id("q"));
    element.sendKeys("selenium");
    Assert.assertEquals("selenium - Google Search", driver.
getTitle());
  }

  @AfterClass
  public void tearDown() throws Exception {
    driver.quit();
  }
}
```

Click on the **Tests** tab to check the log results. The logs are well organized with test steps, screenshots, videos, and a summary. Screenshots are captured at each and step to make the tests more precise, as follows:

```
capabillities.setCapability("screenshot", true); // screenshot
capabillities.setCapability("screenrecorder", true); // video capture
```

TestingBot provides a unique feature by scheduling and running tests directly from the site. The tests can be scheduled to repeat any number of times on a daily or weekly basis. It's even more accurate on scheduling the test start time. You will be apprised of test failures with an alert through e-mail, an API call, an SMS, or a Prowl notification. This feature enables error handling to rerun failed tests automatically as per the user settings.

Launch the Selenium IDE, record tests, and save the test case or test suite in default format (HTML). Access the `https://testingbot.com/` URL from your web browser and click on the **Test Lab** tab. Now, try to upload the already-saved Selenium test case, select the OS platform and browser name and version. Finally, save the settings and execute tests. The test results are recorded and displayed under **Tests**.

Selenium headless browser testing

A headless browser is a web browser without **Graphical User Interface (GUI)**. It accesses and renders web pages but doesn't show them to any human being. A headless browser should be able to parse JavaScript. Currently, most of the systems encourage tests against headless browsers due to their efficiency and time-saving properties. PhantomJS and HTMLUnit are the most commonly-used headless browsers. Capybara-webkit is another efficient headless WebKit for rails-based applications.

PhantomJS

PhantomJS is a headless WebKit scriptable with the JavaScript API. It is generally used for the headless testing of web applications that come with GhostDriver built-in. Tests on PhantomJS are obviously fast since it has fast and native support for various web standards, such as DOM handling, CSS selector, JSON, canvas, and SVG. In general, WebKit is a layout engine that allows web browsers to render web pages. Some browsers, such as Safari and Chrome, use WebKit.

Apparently, PhantomJS is not a test framework, it is a headless browser that is used only to launch tests via a suitable test runner called **GhostDriver**. GhostDriver is a JS implementation of the WebDriver Wire Protocol for PhantomJS; WebDriver Wire Protocol is a standard API that communicates with the browser. By default, GhostDriver is embedded with PhantomJS.

[

To download PhantomJS, refer to `http://phantomjs.org/download.html`.
]

Download PhantomJS, extract the zipped file (for example, `phantomjs-1.x.x-windows.zip` for Windows) and locate the `phantomjs.exe` folder. Add the following imports to your test code:

```
import org.openqa.selenium.phantomjs.PhantomJSDriver;
import org.openqa.selenium.phantomjs.PhantomJSDriverService;
import org.openqa.selenium.remote.DesiredCapabilities;
```

Introduce **PhantomJSDriver** using capabilities to enable or disable JavaScript or to locate the `phantomjs` executable file path:

```
DesiredCapabilities caps = new DesiredCapabilities();
caps.setCapability("takesScreenshot", true);
caps.setJavascriptEnabled(true); // not really needed; JS is enabled
by default
caps.setCapability(PhantomJSDriverService.PHANTOMJS_EXECUTABLE_PATH_
PROPERTY, "C:/phantomjs.exe");
WebDriver driver = new PhantomJSDriver(caps);
```

Alternatively, PhantomJSDriver can also be initialized, as follows:

```
System.setProperty("phantomjs.binary.path", "/phantomjs.exe");
WebDriver driver = new PhantomJSDriver();
```

PhantomJS supports screen capture as well. Since PhantomJS is a WebKit and a real layout and rendering engine, it is feasible to capture a web page as a screenshot. It can be set as follows:

```
caps.setCapability("takesScreenshot", true);
```

The following is the test snippet to capture a screenshot on a test run:

```
File scrFile = ((TakesScreenshot)driver).getScreenshotAs(OutputType.
FILE);
FileUtils.copyFile(scrFile, new File("c:\\sample.jpeg"),true);
```

For example, check the following test program for more details:

```
package packagename;

import java.io.File;
import java.util.concurrent.TimeUnit;
import org.apache.commons.io.FileUtils;
import org.openqa.selenium.*;
import org.openqa.selenium.phantomjs.PhantomJSDriver;

public class phantomjs {
  private WebDriver driver;
  private String baseUrl;

  @BeforeTest
  public void setUp() throws Exception {
    System.setProperty("phantomjs.binary.path", "/phantomjs.exe");
    driver = new PhantomJSDriver();
    baseUrl = "https://www.google.co.in";
    driver.manage().timeouts().implicitlyWait(30, TimeUnit.SECONDS);
  }

  @Test
  public void headlesstest() throws Exception {
  driver.get(baseUrl + "/");
  driver.findElement(By.name("q")).sendKeys("selenium essentials");
  File scrFile = ((TakesScreenshot) driver).
  getScreenshotAs(OutputType.FILE);
  FileUtils.copyFile(scrFile, new File("c:\\screen_shot.jpeg"), true);
  }

  @AfterTest
  public void tearDown() throws Exception {
    driver.quit();
  }
}
```

HTMLUnitDriver

HTMLUnit is a headless (GUI-less) browser written in Java and is typically used for testing. HTMLUnitDriver, which is based on HTMLUnit, is the fastest and most lightweight implementation of WebDriver. It runs tests using a plain HTTP request, which is quicker than launching a browser and executes tests way faster than other drivers. The HTMLUnitDriver is added to the latest Selenium servers (2.35 or above).

The JavaScript engine used by HTMLUnit (Rhino) is unique and different from any other popular browsers available on the market. HTMLUnitDriver supports JavaScript and is platform independent. By default, JavaScript support for HTMLUnitDriver is disabled. Enabling JavaScript in HTMLUnitDriver slows down test execution, but it is advised to enable JavaScript support because most modern sites are Ajax-based web apps. Enabling JavaScript also throws a number of JavaScript warning messages in the console during test execution. The following snippet lets you enable JavaScript for HTMLUnitDriver:

```
HtmlUnitDriver driver = new HtmlUnitDriver();
driver.setJavascriptEnabled(true); // enable JavaScript
```

The following line of code is an alternate way to enable JavaScript:

```
HtmlUnitDriver driver = new HtmlUnitDriver(true);
```

The following piece of code lets you handle a transparent proxy using HTMLUnitDriver:

```
HtmlUnitDriver driver = new HtmlUnitDriver();
driver.setProxy("xxx.xxx.xxx.xxx", port);  // set proxy for handling
Transparent Proxy
driver.setJavascriptEnabled(true);  // enable JavaScript [this emulate
IE's js by default]
```

HTMLUnitDriver can emulate a popular browser's JavaScript in a better way. By default, HTMLUnitDriver emulates IE's JavaScript. To handle the Firefox web browser with version 17, use the following snippet:

```
HtmlUnitDriver driver = new HtmlUnitDriver(BrowserVersion.FIREFOX_17);
driver.setJavascriptEnabled(true);

Here is the snippet to emulate a specific browser's JavaScript using
capabilities:
DesiredCapabilities capabilities = DesiredCapabilities.htmlUnit();
driver = new HtmlUnitDriver(capabilities);
```

```
DesiredCapabilities capabilities = DesiredCapabilities.firefox();
capabilities.setBrowserName("Mozilla/5.0 (X11; Linux x86_64; rv:24.0)
Gecko/20100101 Firefox/24.0");
capabilities.setVersion("24.0");
driver = new HtmlUnitDriver(capabilities);
```

Switching user agents

Earlier, cryptic commands and texts were used to retrieve data from the Internet, which would act as a user agent; now, web browsers are used as user agents. To put it simply, a user agent is a tool to browse the Internet by faking another browser. In general, websites are being rendered differently on different browsers with different platforms (for example, Chrome browser on Windows 7). However, the user agents cannot render a web page similar to the selected one. The user can track their own user agent string from http://whatsmyuseragent.com/.

For example, Mozilla Firefox 5.0 (Windows NT 6.1) AppleWebKit/537.36 (KHTML, like Gecko) Chrome/36.0.1985.143 Safari/537.36.

Here, **Mozilla/5.0** is an application name and **Chrome/36.0.1985.143** is the browser type, with version. Whenever the user enters a URL in the web browser, a request will be sent to the web server to identify the user agent; it provides the relevant data as response. Selenium WebDriver has extended its features to drive tests over different user agents and execute tests from browser profiles.

Firefox user agent

Follow these steps to run tests on your favorite Firefox user agents:

1. Launch Firefox and install User Agent Switcher, the Firefox add-on.
2. From the **Tools** menu, navigate to **Tools** | **Default User Agent** | **Edit User Agents** and click on **Edit User Agents**.
3. Select any one of the user agents from the list (for example, **Googlebot 2.1**) and click on the **Edit** button.
4. Now, get the user agent string, as follows:

   ```
   Mozilla/5.0 (compatible; Googlebot/2.1; +http://www.google.com/
   bot.html)
   ```

Make use of the above user agent string to run tests through Googlebot 2.1. To run tests on a specific version of Firefox, refer to `http://www.useragentstring.com/pages/Firefox/`. The following code encapsulates the discussion in this paragraph:

```
ProfilesIni profile = new ProfilesIni();
FirefoxProfilemyprofile = profile.getProfile("default");
myprofile.setPreference("general.useragent.override", "Mozilla/5.0
(compatible; Yahoo! Slurp; http://help.yahoo.com/help/us/ysearch/
slurp)");  // here, the user-agent is 'Yahoo Slurp'
WebDriver driver = new FirefoxDriver(myprofile);
```

Chrome user agent

Follow these steps to run tests on your favorite Chrome user agents:

1. Launch Google Chrome and install the Chrome add-on User-Agent Switcher for Chrome.

2. Go to `chrome://extensions/` in the Chrome web browser and click on options `link_text`.

3. Obtain the desired user agent string (for example, **iPhone 4**). The following is the stream for **iPhone4**:

   ```
   Mozilla/5.0 (iPhone; U; CPU iPhone OS 4_3_2 like Mac OS X; en-us)
   AppleWebKit/533.17.9 (KHTML, like Gecko) Version/5.0.2 Mobile/8H7
   Safari/6533.18.5
   ```

   The following screenshot shows us the **Extensions** page in which **User-Agent Switcher for Chrome** is added:

Make use of the obtained user agent string to run the tests through the iPhone 4 user agent from Chrome, as follows:

```
System.setProperty("webdriver.chrome.driver","C:\\chromedriver.exe");
ChromeOptions options = new ChromeOptions();
options.addArguments("user-data-dir=C:/Users/user_name/AppData/Local/
Google/Chrome/User Data");
```

```
options.addArguments("--user-agent=Mozilla/5.0 (iPhone; U; CPU iPhone
OS 4_3_2 like Mac OS X; en-us) AppleWebKit/533.17.9 (KHTML, like
Gecko) Version/5.0.2 Mobile/8H7 Safari/6533.18.5"); //iPhone 4
options.addArguments("--start-maximized");
driver = new ChromeDriver(options);
```

Tests on specific Firefox versions

The Firefox binary lets you run tests on your favorite Firefox versions. In order to do that, perform the following steps:

1. Install multiple versions of Firefox on your PC (say, FF 26 and FF 28). Make sure that the Mozilla Firefox versions are installed at different path locations using custom installation.

2. Add the following imports in your test code:

```
import java.io.File;
import org.openqa.selenium.firefox.FirefoxBinary;
import org.openqa.selenium.firefox.FirefoxDriver;
import org.openqa.selenium.firefox.FirefoxProfile;
```

3. Locate the secondary Firefox executable path in the Firefox binary.

4. Create a Firefox profile and initialize WebDriver as shown in the following code snippet:

```
FirefoxBinary binary = new FirefoxBinary(new File("C://Program
Files//Mozilla Firefox26//firefox.exe"));
FirefoxProfile profile = new FirefoxProfile();
WebDriver driver = new FirefoxDriver(binary, profile);
```

5. For Python bindings, use a similar scenario and add the following code snippet:

```
from selenium.webdriver.firefox.firefox_binary import
FirefoxBinary
driver = webdriver.Firefox
(firefox_binary=FirefoxBinary("C://Program Files//Mozilla
Firefox26//firefox.exe"))
```

Tests from the custom Firefox profile

In general, custom profiles are used in order to get rid of control over existing cookies that contain history, bookmarks, passwords, personal information, and so on.

Firefox Profile Manager is used to create or remove Firefox profiles. To create a Firefox profile, perform the following steps:

1. Open the **Run** command window (using Windows Key + *R*), type `firefox.exe -p` and click on **OK**.

 Mozilla Firefox must be closed before opening the Firefox Profile Manager. If the Profile Manager window does not open, Firefox must be running in the background. Close all instances of Firefox, restart the computer, and then try again to solve the issue.

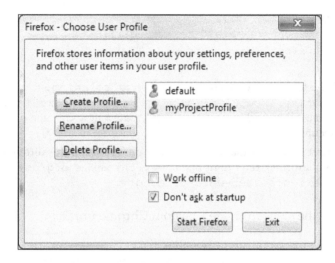

2. Create a Firefox profile (for example, `myProjectProfile`) by clicking on the **Create Profile** button from the profile manager.

3. Add the following snippet in your test code to run tests from the new, customized Firefox profile:

```
ProfilesIni profile = new ProfilesIni();
FirefoxProfilemyprofile = profile.getProfile("myProjectProfile");
WebDriver driver = new FirefoxDriver(myprofile);
```

In Linux, Firefox native events are disabled by default as they may launch more than one browser in parallel in a test. To enable such default Firefox-disabled features, the native events insist on being set to `true`, as shown in the following code:

```
FirefoxProfile profile = new FirefoxProfile();
profile.setEnableNativeEvents(true);
WebDriver driver = new FirefoxDriver(profile);
```

Tests from the custom Chrome profile

Google Chrome is always the preferable browser on which to run tests. The Firefox and Chrome browsers work with binary. However, Internet Explorer doesn't need a profile to be set up to run tests because they run on the server user. In Chrome, the user data directory contains all the data of the given user, which includes history, bookmarks, and cookies. Add the following snippet to run tests from a new profile:

```
System.setProperty
("webdriver.chrome.driver","C:\\chromedriver.exe");
ChromeOptions options = new ChromeOptions();
options.addArguments("user-data-dir=C:/Users/user_name/AppData/Local/
Google/Chrome/User Data");
options.addArguments("--start-maximized");
driver = new ChromeDriver(options);
```

If you face an error such as org.openqa.selenium.
WebDriverException: unknown error: Chrome failed to
start: exited normally, then create a new Chrome profile and
execute the tests.

To get more information on the user data location (for different
OS platforms), refer to http://www.chromium.org/user-
experience/user-data-directory.

The following are the steps to create a custom Chrome profile:

1. Copy the User Data folder (located at C:/Users/user_name/AppData/
   Local/Google/Chrome/User Data) and paste it in the same folder with
   a different name (for example, New User).

2. Open the New User folder.

3. Rename the Default folder, so that after your test run, a new folder named
   Default is created.

4. Now, replace the directory in your code with C:/Users/user_name/
   AppData/Local/Google/Chrome/New User.

5. To verify the new profile, try to bookmark some of the websites
   and observe them in the next run.

Summary

In this chapter, we have learned how to perform Selenium cross-browser automation tests, how to run automation tests in the cloud, and how to switch user agents.

In the next chapter, we will discuss all about the functions of Selenium WebDriver and its uses.

3
Selenium WebDriver Functions

Selenium WebDriver API provides an object-oriented approach to test web-based applications. On the one hand, Selenium RC (Remote Control) injects JavaScript into the browser on runtime, and on the other hand, Selenium WebDriver performs direct calls to the browser using each browser's native approach. Meanwhile, WebDriverBackedSelenium lets you combine both Selenium RC and WebDriver. Hopefully, Selenium RC will be deprecated and taken away in Selenium 3.0, which includes the Marionette driver replacing the Firefox driver to automate Firefox OS in mobile platforms. WebDriver API enriched Selenium RC is called Selenium WebDriver or Selenium 2.0. In the future, Selenium will meet all the W3C standards; however, the current Selenium WebDriver API encloses a bunch of functions used for effective web automation tests.

In this chapter, we will learn about the following topics:

- Basic Selenium WebDriver functions
- Locating WebElements
- WebElement and Windows WebDriver functions
- Navigation and Cookies WebDriver functions
- Select functions
- Handling alerts and pop-ups
- Mouse and keyboard actions

Basic WebDriver functions

Consider a test use case with the user opening a browser, searching for a term, asserting the actual value with the expected value, and then finally exiting the browser. This simple use case can certainly be achieved using the basic Selenium WebDriver functions. The elementary functions, such as `click()`, `close()`, `submit()`, and `sendKeys()`, are the fundamental keys to start with any test automation tasks. Let's discuss the basic WebDriver functions in more detail:

- `close()`: This function exits or closes the current active browser window. The following is the syntax for this function:

  ```
  driver.close();
  ```

- `quit()`: This function halts the running driver and closes every browser window by ending the active session. The following is the syntax for this function:

  ```
  driver.quit();
  ```

- `getTitle()`: This function fetches the current page title. The following is the syntax for this function:

  ```
  driver.getTitle();
  ```

- `getCurrentUrl()`: This function gets the current web page URL. The following is the syntax for this function:

  ```
  driver.getCurrentUrl();
  ```

- `getPageSource()`: This function retrieves the entire page source of the loaded web page and allows the user to assert any text present in the same web page. However, the modified DOM, due to asynchronous (Ajax) calls, is not reflected on some of the browsers. Instead, it returns the page source of the previously loaded web page. The `getPageSource()` method is not advisable for web pages loading JavaScripts asynchronously. The following is the syntax for this function:

  ```
  driver.getPageSource();
  ```

  Some of the helpful snippets using this function are given as follows:

  - Verify text:

    ```
    driver.getPageSource().contains("your text");
    ```

  - Assert text:

    ```
    boolean b = driver.getPageSource().contains("your text");
    System.out.println(b);
    assertTrue(b);
    ```

- `click()`: This function lets you click on a link, button, checkbox, or radio button. The following is the syntax for this function:

```
driver.findElement(By.locatorType("path")).click();
```

The following is an example that covers the `click()` function to do a simple Google search on clicking the search button:

```
driver.get("https://www.google.com");
driver.findElement(By.name("q")).sendKeys("selenium essentials");
driver.findElement(By.id("sblsbb")).click();
```

- `clear()`: This function erases or empties the string values present in a text field. The following is the syntax for this function:

```
driver.findElement(By.locatorType("path")).clear();
```

The following code snippet is a simple Google search, where the `clear()` method is used to erase the current search keyword from the search text field and prepare for the next search:

```
driver.get("https://www.google.com");
driver.findElement(By.name("q")).sendKeys("selenium essentials");
driver.findElement(By.name("q")).clear();
driver.findElement(By.name("q")).sendKeys("prashanth sams");
```

- `sendKeys()`: This function lets you type or insert text in the text field on runtime. The following is the syntax for this function:

```
driver.findElement(By.locatorType("path")).sendKeys("your text");
```

The following snippet explains the `sendKeys()` function that lets you insert text or a sentence in the Google search text field:

```
driver.get("https://www.google.com");
WebElement element = driver.findElement(By.name("q"));
element.sendKeys("selenium essentials");
```

- `submit()`: This function is similar to the `click()` function. However, this function is used to submit a form with `<form>` tags. The following is the syntax for this function:

```
driver.findElement(By.locatorType("path")).submit();
```

The `submit()` function is commonly used instead of pressing the *Enter* key. Let's see an example to retrieve results from the Google search:

```
driver.get("https://www.google.com");
WebElement element = driver.findElement(By.name("q"));
element.sendKeys("selenium essentials");
element.submit();
```

Locating WebElements

Element-locating functions are the building blocks of Selenium tests. These methods handle Ajax calls using timeouts and wait conditions to search and locate elements within a web page. There is no fixed strategy that you can follow to locate an element; it depends on the user ideology and their comfort level in locating the web elements.

An element can be located using any kind of locator type. Selenium WebDriver uses locators to interact with elements present in a web page. The following is the list of locator types used in Selenium WebDriver to locate web elements:

- `By.id`
- `By.name`
- `By.xpath`
- `By.cssSelector`
- `By.className`
- `By.linkText`
- `By.tagName`
- `By.partialLinkText`

The following is the syntax to use `locatorType`:

```
driver.findElement(By.locatorType("path"))
```

The following is an example for `locatorType`:

```
driver.findElement(By.id("sblsbb")).click();
```

Prioritize the locator type in this order: `id` > `name` > `css` > `xpath`. The `cssSelector` locator type is a good pick to work with Ajax calls. Always try to avoid locating elements using XPath locators. Remember that Internet Explorer tests often fail to respond to XPath locators. The XPath locator type is of two categories, namely, absolute XPath `/` and relative XPath `//`. Consider an element on the Google search page:

Absolute XPath	Relative XPath
`//html/body/div[1]/div[3]/form/div[2]/` `div[2]/div[1]/div[1]/div[3]/div[1]/div[3]/` `div[1]/input[1]`	`//input[@id='lst-` `ib']`

Here, both the absolute and relative XPath point to the same element.

 To know more on XPath functions and axes, please refer to the
following links:
- `http://bit.ly/ZSBNzp`
- `http://bit.ly/1w7YIna`

Let's take a look at Selenium WebDriver's element-locating functions, which are
as follows:

- `findElement()`: This element locates the first element within the current
  page. The following is the syntax for this function:

```
driver.findElement(By.locatorType("path"));
```

  The following is an example where Selenium WebDriver locates the Google
  search text field using the `findElement()` method:

```
driver.get("https://www.google.com");
WebElement element = driver.findElement(By.id("lst-ib"));
```

- `findElements()`: This element locates all the elements within the current
  page. The following is the syntax for this function:

```
List<WebElement> elements = driver.findElements(By.
locatorType("path"));
WebElement element = driver.findElement(By.locatorType("path"));
List<WebElement> elements = element.findElements(By.
locatorType("path"));
```

Some of the useful tasks that you can perform with the help of these functions are
as follows:

- **Store Length**: The following code snippet captures the list of textboxes
  available in a Google authentication page:

```
driver.get("https://accounts.google.com/");
List<WebElement> Textbox =driver.findElements(By.xpath("//script[@
type='text/javascript']"));
System.out.println("Overall textboxes:"+Textbox.size());
```

- **Click Element**: Let's go through a book search on clicking on any one of the
  available autosuggestions. Fortunately, the first option from the list is chosen,
  which returns contents related to it as the search result. The following code
  snippet does the work for us:

```
driver.get("http://www.indiabookstore.net");
driver.findElement(By.id("searchBox")).sendKeys("Alche");
```

```
List <WebElement> listItems = driver.findElements(By.xpath("//
div[3]/ul/li"));
listItems.get(0).click();
```

- **Locate by tag name**: Capturing elements using tag names is widely used to collect autosuggestions, checkboxes, ordered lists, and so on. The following snippet is an example for capturing elements through the `tagName` function:

```
List<WebElement> link = driver.findElement(By.
locatorType("path")).findElements(By.tagName("li"));
```

- **Multi-select elements**: This snippet gives you a clear idea on how to use the multi-select functionality. Here, the `findElements()` method is used to return the total length of the option tag, and it lets you pick all the preferred checkboxes:

```
driver.get("http://www.ryancramer.com/journal/entries/select_
multiple/");
List<WebElement> ele = driver.findElements(By.tagName("select"));
System.out.println(ele.size());
WebElement ele2 = ele.get(0);
List<WebElement> ele3 = ele2.findElements(By.tagName("option"));
System.out.println(ele3.size());
ele2.sendKeys(Keys.CONTROL);
ele3.get(0).click();
ele3.get(1).click();
ele3.get(3).click();
ele3.get(4).click();
ele3.get(5).click();
```

- **Capture and navigate all links in a web page**: As discussed in the preceding section, the element list is captured and stored in a single dimensional array using the `findElements()` method. Later, the user is intended to navigate each and every link one by one. The following snippet is an example for capturing and navigating all links in a web page:

```
private static String[] links = null;
private static int linksCount = 0;
driver.get("https://www.google.co.in");
List<WebElement> linksize = driver.findElements(By.tagName("a"));
linksCount = linksize.size();
System.out.println("Total no of links Available: "+linksCount);
links= new String[linksCount];
System.out.println("List of links Available: ");
// print all the links from webpage
for(int i=0;i<linksCount;i++)
{
```

```
    links[i] = linksize.get(i).getAttribute("href");
}
// navigate to each Link on the webpage
for(int i=0;i<linksCount;i++)
{
  driver.navigate().to(links[i]);
  Thread.sleep(3000);
}
```

- **Capture all links under specific frame/class/ID and navigate one by one**:
  In general, `StaleElementException` will be thrown whenever a user
  navigates through the link and tries to click on the second link after returning
  from the visited page. To avoid such exceptions, an external method,
  `getElementWithIndex()`, is used to return values. The following code
  snippet exemplifies this method:

```
driver.get("https://www.google.co.in");
WebElement element = driver.findElement(By.locatorType("path"));
List<WebElement> elements = element.findElements(By.tagName("a"));
int sizeOfAllLinks = elements.size();
System.out.println(sizeOfAllLinks);
for(int i=0; i<sizeOfAllLinks ;i++)
{
  System.out.println(elements.get(i).getAttribute("href"));
}
for (int index=0; index<sizeOfAllLinks; index++ ) {
  getElementWithIndex(By.tagName("a"), index).click();
  driver.navigate().back();
}

public WebElement getElementWithIndex(By by, int index) {
  WebElement element = driver.findElement(By.id(Value));
  List<WebElement> elements = element.findElements(By.
tagName("a"));
  return elements.get(index);
}
```

- **Capture all links**: The following model follows looping conditions to capture
  all the links present in a web page:

```
driver.get("https://www.google.co.in");
List<WebElement> all_links_webpage = driver.findElements(By.
tagName("a"));
System.out.println("Total no of links Available: " + all_links_
webpage.size());
int k = all_links_webpage.size();
```

```
System.out.println("List of links Available: ");
for(int i=0;i<k;i++)
{
  if(all_links_webpage.get(i).getAttribute("href")
    .contains("google"))
  {
    String link = all_links_webpage.get(i)
      .getAttribute("href");
    System.out.println(link);
  }
}
```

- **Locate and select autocomplete**: The following snippet captures and clicks on a list of autocomplete results grown at runtime using the list `<li>` tag:

```
driver.get("http://www.indiabookstore.net");
driver.findElement(By.id("searchBox")).sendKeys("Alche");
Thread.sleep(3000);
List <WebElement> listItems = driver.findElements(By.
cssSelector(".acResults li"));
listItems.get(0).click();
driver.findElement(By.id("searchButton")).click();
```

- **Store autosuggestions using iterator**: The iterator is an alternative method to handle loops. In this example, a list of autosuggestions is iterated and stored:

```
driver.get("http://www.indiabookstore.net/");
driver.findElement(By.id("searchBox")).sendKeys("sam");
WebElement table = driver.findElement(By.className("acResults"));
List<WebElement> rowlist = table.findElements(By.tagName("li"));
System.out.println("Total No. of list: "+rowlist.size());
Iterator<WebElement> i = rowlist.iterator();
System.out.println("Storing Auto-suggest..........");
while(i.hasNext())
{
  WebElement element = i.next();
  System.out.println(element.getText());
}
```

- The looping iteration can also be wrapped with the following snippet. It is as simple as the preceding snippet. Here, the list items are declared to a variable named called `temp`. Through `temp`, the values are being retrieved:

```
List<WebElement> listitems = driver.findElements(By.id("value"));
for(WebElement temp: listitems) // temp is the declared variable
name
{
```

```
System.out.println
((temp.findElement(By.tagName("value")).getText()));
}
```

WebElement functions

WebElement is an HTML element that helps the users to drive automation tests. Selenium WebDriver provides well-organized web page interactions through WebElements, such as locating elements, getting attribute properties, asserting text present in WebElement, and more. However, to interact with hidden elements in a web page, it is necessary to unhide the hidden elements first. Let's discuss Selenium WebElement functions further:

- `getText()`: This function delivers the `innerText` attribute of WebElement. The following is the syntax for this function:

  ```
  driver.findElement(By.locatorType("path")).getText();
  ```

  The following is an example on the Google page that returns the `innerText` attribute of a Google search button using the `getText()` function:

  ```
  driver.get("https://www.google.com");
  System.out.println(driver.findElement(By.id("_eEe")).getText());
  ```

  `JavaScriptExecutor` is a Selenium interface to execute JavaScripts that returns the `innerText` attribute of a hidden element. It is important to unhide the hidden elements before extracting `innerText`, as shown in the following code snippet:

  ```
  WebElement Element = driver.findElement(By.locatorType("path"));
  JavascriptExecutor jse = (JavascriptExecutor)driver;
  System.out.println(jse.executeScript("return arguments[0].
  innerHTML", Element));
  ```

- `getAttribute()`: This function delivers the value of a given attribute of an element. The following is the syntax for this function:

  ```
  driver.findElement(By.locatorType("path")).getAttribute("value");
  ```

  The values or properties of an attribute can be easily returned using the `getAttribute()` method. The attributes can be a `class/id/name/value/` or any other attribute. The following piece of code returns the class attribute value of a Google search button:

  ```
  driver.get("https://www.google.com");
  driver.findElement(By.xpath("//div[@id='lst-ib']")).
  getAttribute("class");
  ```

- `getTagName()`: This function delivers the tag name of a given element. The following is the syntax for this function:

```
driver.findElement(By.locatorType("path")).getTagName();
```

Let's see a snippet to get `tagName` of a Google search text field element using the `getTagName()` function:

```
driver.get("https://www.google.com");
driver.findElement(By.xpath("//div[@class='sbib_b']")).
getTagName();
```

- `isDisplayed()`: This function checks whether an element is displayed in a page or not. It returns a Boolean value (`true` or `false`). The following is the syntax for this function:

```
driver.findElement(By.locatorType("path")).isDisplayed();
```

This method confirms whether an element is visible in a page or not until the timeout occurs. The Google search text field element is asserted here to acknowledge that the page is opened properly or not. The following code snippet does the work for us:

```
driver.get("https://www.google.com");
WebElement Element = driver.findElement(By.name("q"));
Assert.assertTrue(Element.isDisplayed());
```

- `isEnabled()`: This function checks whether an element is enabled in a page or not. It returns a Boolean value (`true` or `false`). The following is the syntax for this function:

```
driver.findElement(By.locatorType("path")).isEnabled();
```

Let's see an example that checks the element's status using the `isEnabled()` method. In general, the user is not allowed to edit the text field when an element is disabled. However, if it is in the enabled status, the right action has to be performed, or an assertion failure can be thrown as follows:

```
driver.get("https://www.google.com");
WebElement Element = driver.findElement(By.name("q"));
if(Element.isEnabled())
{
   driver.findElement(By.name("q")).sendKeys("Selenium
Essentials");
}else{
   Assert.fail();
}
```

- `isSelected()`: This function verifies whether an element is selected or not. It returns a Boolean value (`true` or `false`). The following is the syntax for this function:

```
driver.findElement(By.locatorType("path")).isSelected();
```

In the following example, the `if` condition is used to confirm whether the combobox is selected or not:

```
driver.get("http://www.angelfire.com/fl5/html-tutorial/ddmenu.
htm");
WebElement Element1 = driver.findElement(By.xpath("//select[@
name='jump']/option[1]"));
WebElement Element2 = driver.findElement(By.xpath("//select[@
name='jump']/option[2]"));
if(Element1.isSelected())
{
  System.out.println("html tutorial is selected");
}else if (Element2.isSelected()){
  System.out.println("altavista is selected");
}
```

- `getSize()`: This method returns the width and height (dimensions) of a rendered element, as follows:

```
Dimension dimensions=driver.findElement(By.locatorType("path")).
getSize();
dimensions.width;
dimensions.height;
```

Let's see how we can return the width and height of a Google logo in the Google search page:

```
driver.get("https://www.google.com");
System.out.println(driver.findElement(By.xpath("//div[@
id='hplogo']")).getSize());
Dimension dimensions=driver.findElement(By.xpath("//div[@
id='hplogo']")).getSize();
System.out.println("Logo Width : "+dimensions.width);
System.out.println("Logo Height : "+dimensions.height);
```

- `getLocation()`: This function returns the *x* and *y* coordinates with the point location of the top-left corner of an element.

```
Point point = driver.findElement(By.locatorType("path")).
getLocation();
point.x;
point.y;
point.getX();
point.getY();
```

The `getLocation()` method returns the values as point objects. Let's see how to return the *x* and *y* coordinates of a Google logo in the Google search page. The following are the two methods to use the `getLocation()` method:

- Method 1:

```
driver.get("https://www.google.com");
Point point = driver.findElement(By.xpath("//div[@
id='hplogo']")).getLocation();
System.out.println("X Position : " + point.x);
System.out.println("Y Position : " + point.y);
```

- Method 2:

```
driver.get("https://www.google.com");
Point point = driver.findElement(By.xpath("//div[@
id='hplogo']")).getLocation();
System.out.println(point);
System.out.println(point.getX() + "\t" + point.getY());
```

- `getCssValue()`: This method returns the value of any CSS properties. The following is the syntax for this function:

```
driver.findElement(By.locatorType("path")).getCssValue("font-
size");
```

WebElement's style attribute values can be certainly attained using this method. Let's focus on the CSS properties of the Google logo, as follows:

```
driver.get("https://www.google.com");
WebElement element = driver.findElement(By.xpath("//div[@
id='hplogo']"));
System.out.println(element.getCssValue("font-size"));
System.out.println(element.getCssValue("font-weight"));
System.out.println(element.getCssValue("color"));
System.out.println(element.getCssValue("background-size"));
```

Navigation

Page navigation kicks in from the start of test execution; it is a fundamental task for each and every use case, especially to browse through the history and navigate backwards and forward. Let's discuss Selenium navigation functions, as follows:

- The `get()` function commands the browser to navigate to the URL. In general, the `get()` function is used to open a web page on every test execution. The `onload` event lets the browser wait until the complete page is loaded. If the page is overloaded with lots of Ajax calls, there will be delays on page load. The following is the syntax for this function:

  ```
  driver.get("URL");
  ```

- The `navigate().back()` function lets the browser navigate backwards. The following is the syntax for this function:

  ```
  driver.navigate().back();
  ```

  Here is an alternative method using keyboard actions to navigate web browser history.

  ```
  Actions actions = new Actions(driver);
  actions.sendKeys(Keys.BACK_SPACE).perform();
  ```

- The `navigate().forward()` function allows the browser to navigate forward. This function lets you move a page forward, which is similar to clicking a browser's forward button. The following is the syntax for this function:

  ```
  driver.navigate().forward();
  ```

- The `navigate().to()` function is similar to the `get()` function in order to access a web page. This function is normally used whenever a user needs to navigate to a specific URL at test. The following is the syntax for this function:

  ```
  driver.navigate().to("URL");
  ```

- The `refresh()` method refreshes the current web page. A page refresh is nothing but reloading a full page. The following is the syntax of this function:

  ```
  driver.navigate().refresh();
  ```

There are numerous ways to refresh the current web page. Let's see the methods one by one; however, a few methods are not applicable to Mac:

Method 1:

```
Actions actions = new Actions(driver);
actions.keyDown(Keys.CONTROL).sendKeys(Keys.F5).
perform();
```

Method 2:

```
JavascriptExecutor js = (JavascriptExecutor) driver;
js.executeScript("history.go(0)");
```

Method 3:

```
driver.navigate().to(driver.getCurrentUrl());
```

Method 4:

```
driver.findElement(By.locatorType("path")).sendKeys("\
uE035");
```

Method 5:

```
driver.findElement(By.locatorType("path")).sendKeys(Keys.
F5);
```

Cookies

Cookies are generally stored in a web browser to identify the user's activity when browsing a web page. Selenium WebDriver handles web browser cookies by finding, deleting, modifying, and adding the cookies. It provides an excellent mechanism to modify cookies for high-level automation testing. Let's discuss Selenium WebDriver cookies, as follows:

* The getCookies() method returns cookies present in a loaded page. The following is the syntax for this function:

```
driver.manage().getCookies();
```

Some of the useful tasks that you can perform with the help of getCookies() function are discussed as follows:

> **Return cookies**: The getCookies() method captures all the cookies generated in a web page. Let's see a model to get to know how we can return the entire cookie ID and expiry date and time and the domain name of each and every cookie:
>
> ```
> driver.get("https://www.google.co.in");
> System.out.println(driver.manage().getCookies());
> ```

- The following piece of code lets you explain how to get all the cookies in a systematic manner:

```
driver.get("https://www.google.co.in");
Set<Cookie> cookies = driver.manage().getCookies();
for(Cookie ck :cookies) {
  System.out.println(ck);
}
```

- The `getCookieNamed()` returns a specific cookie by name. The following is the syntax for this function:

```
driver.manage().getCookieNamed(String arg0);
```

The following snippet returns cookies with respect to the name from the cookie file (memory of the browser); here, the generated cookie named NID is reached from the Google page:

```
driver.get("https://www.google.co.in");
System.out.println(driver.manage().getCookies());
System.out.println(driver.manage().getCookieNamed("NID"));
```

- The `addCookie()` method injects user-defined cookies into the loading page at runtime:

```
driver.manage().addCookie(Cookie arg0);
```

Adding a cookie replaces the existing cookie if one exists. Let's see how we can inject a new cookie into the current domain using the following code snippet:

```
driver.get("https://www.google.co.in");
System.out.println(driver.manage().getCookies());
Cookie cookie = new Cookie("NID",
"67=h88c3NpCuTQABjgZF2Ix8CJHivtpYDLFk5gc1_dtEnz1aP_
UugPSWGukXUPeKPXOeTKZdkcWrw-DnqjsOEGhL7sURlkhamIAxsBUWH_
Hh76MQ490jfT9pdwsMkWoYJAJ");
driver.manage().addCookie(cookie);
System.out.println("-----------------------");
driver.get("https://www.google.co.in");
System.out.println(driver.manage().getCookies());
```

- The `deleteCookie()` method deletes the stored cookie in the loaded webpage. The following is the syntax for this function:

```
driver.manage().deleteCookie(Cookie arg0);
```

Deleting a cookie or a set of cookies means erasing certain cookies from the browser's `cookie` file. The following snippet explains the `deleteCookie()` method in different techniques:

```
driver.get("https://www.google.co.in");
Set<Cookie> cookies =  driver.manage().getCookies();
for(Cookie ck :cookies) {
        driver.manage().deleteCookie(ck);
}
System.out.println(driver.manage().getCookies());
driver.get(driver.getCurrentUrl());
Cookie cookie = new Cookie("NID", "67=QKDjS3SxgW9NTe4m
ymVk8t0V_7314Tf1JFtkiVfb27REyOJJfW8NXzCWsandTyCYVllSK-
EO7Vol2yJH1Xam4HmbKmUm7Pvm8g44dtOdm-wfW evNWKRr_UlF3Z34n28e");
driver.manage().deleteCookie(cookie);
```

- The `deleteAllCookies()` method deletes all the stored cookies in a loaded web page. The following is the syntax of this function:

```
driver.manage().deleteAllCookies();
```

Let's delete all the cookies generated and stored in a web page (the current domain) using the following code snippet:

```
driver.get("https://www.google.co.in");
System.out.println(driver.manage().getCookies());
driver.manage().deleteAllCookies();
System.out.println(driver.manage().getCookies());
```

- The `deleteCookieNamed()` method deletes a specific cookie by name. The following is the syntax of this function:

```
driver.manage().deleteCookieNamed(String arg0);
```

This method clears the cookies with respect to the name from the `cookie` file (the memory of the browser). As cookies are stored as name-value pairs, removing a cookie from the current domain is simple using Selenium WebDriver's cookie functions. Let's have a quick glance at how to remove a single cookie from the Google page using the cookie name:

```
driver.get("https://www.google.co.in");
System.out.println(driver.manage().getCookies());
driver.manage().deleteCookieNamed("NID");
System.out.println(driver.manage().getCookies());
```

Window functions

The browser window is a source key for web automation tests. A web page should cover W3C standards and meet responsive design needs. In general, the test status (pass/fail) varies with the varying size of browser windows for a buggy site built without proper responsive design and W3C standards. Selenium WebDriver handles browser window tasks, such as resizing the window, locating the window, and switching windows with its extensive API support without any JavaScript injections from outside. Let's discuss Selenium window functions:

- The `maximize()` method lets you maximize the current browser window size. This function directly responds to the browser window; however, the user needs the `java.awt.Robot` library for Java bindings to work along with the screen resolutions. The following is the syntax of this function:

```
driver.manage().window().maximize();
```

  Maximizing the browser window gives a better view of test execution. By default, the browser window is in normal mode and not maximized. Let's see how to maximize a Firefox browser window using the following code snippet:

```
WebDriver driver = new FirefoxDriver();
driver.manage().window().maximize();
```

  Still, there are a number of ways to maximize a browser window. Some of the methods are cited as snippets, as follows:

  - **Method 1**: In this method, we use the `Robot` class to get the total screen size. Finally, the width and height of the screen are applied to maximize the window:

```
import org.openqa.selenium.Dimension;
import org.openqa.selenium.Point;
WebDriver driver = new FirefoxDriver();
driver.manage().window().setPosition(new Point(0,0));
java.awt.Dimension capturedScreenSize = java.awt.Toolkit.
getDefaultToolkit().getScreenSize();
Dimension d = new Dimension((int) capturedScreenSize.
getWidth(), (int) capturedScreenSize.getHeight());
driver.manage().window().setSize(d);
```

- ○ **Method 2**: Here, the size of the current browser is forced to be set to a fixed `Dimension` with the help of Selenium WebDriver windows functions, as follows:

```
import org.openqa.selenium.Dimension;
import org.openqa.selenium.Point;
WebDriver driver = new FirefoxDriver();
driver.manage().window().setPosition(new Point(0,0));
driver.manage().window().setSize(new
Dimension(1600,768));
```

- ○ **Method 3**: JavaScript is an optional mode to maximize the browser window. The following is the JavaScript to maximize the browser window:

```
JavascriptExecutor jse = (JavascriptExecutor)driver;
jse.executeScript("window.open('','Test','width=200,heig
ht=200')");
driver.close();
driver.switchTo().window("Test");
jse.executeScript("window.moveTo(0,0);");
jse.executeScript("window.resizeTo(1800,800);");
```

- The `getSize()` method stores the width and height of a browser window, an image, or WebElement. The following is the syntax for this function:

```
driver.manage().window().getSize();
```

This method returns dimension values as objects. Here, we go to get the size of the current browser window:

```
driver.get("https://www.google.co.in");
System.out.println(driver.manage().window().getSize());
```

- The `getPosition()` method stores the *x* and *y* coordinates of the top-left corner of the browser window. The following is the syntax of this function:

```
driver.manage().window().getPosition();
```

This method returns the values as point objects. Here, we go to get the position of the current browser window:

```
System.out.println(driver.manage().window().getPosition());
System.out.println("Position X: " + driver.manage().window().
getPosition().x);
System.out.println("Position Y: " + driver.manage().window().
getPosition().y);
System.out.println("Position X: " + driver.manage().window().
getPosition().getX());
```

```
System.out.println("Position Y: " + driver.manage().window().
getPosition().getY());
```

- The `setSize()` method customizes the browser window dimension. The following is the syntax for this function:

```
driver.manage().window().setSize(new Dimension(width, height));
```

Due to the increasing number of mobile users, responsive design is equally in demand to meet industry standards. Selenium offers a better solution to test these responsive sites. Let's look through a short code to customize the browser window:

```
Dimension d = new Dimension(320, 480);
driver.manage().window().setSize(d);
driver.manage().window().setSize(new Dimension(320, 480));
```

- The `setPosition()` method customizes *x* and *y* coordinates of the top-left corner of the browser window:

```
driver.manage().window().setPosition(new Point(X-axis, Y-axis));
```

Screens with varying resolution need attention before locating the browser under test. This method customizes the browser window location using the `Point` object, as follows:

```
Point p = new Point(200, 200);
driver.manage().window().setPosition(p);
driver.manage().window().setPosition(new Point(300, 150));
```

- The `getWindowHandle()` method handles the current browser window (for example, `main | parent window`).

This method lets you handle a browser after switching a specific window being tested. The following is a sample snippet to handle the browser window:

```
String parentwindow = driver.getWindowHandle();
driver.switchTo().window(parentwindow);
```

- The `getWindowHandles()` method handles all the browser windows and allows the user to switch control between the parent window and the child windows. The following is the syntax for this function:

```
driver.getWindowHandles();
```

A browser window can populate any number of child windows. To handle these windows, lists of objects are used to get through them (child windows) one by one. The following is a sample code snippet for this function:

```
Set<String> childwindows =  driver.getWindowHandles();
driver.switchTo().window(childwindow);
```

- The `switchTo.window()` method transfers control from one browser to another browser window. In general, it helps you to switch control from a parent window to a child window and then get back the control to the parent window. The following is the syntax for this function:

```
driver.SwitchTo().Window(childwindow);
driver.close();
driver.SwitchTo().Window(parentWindow);
```

The following image is a pictorial representation of a browser with multiple windows (parent and child).

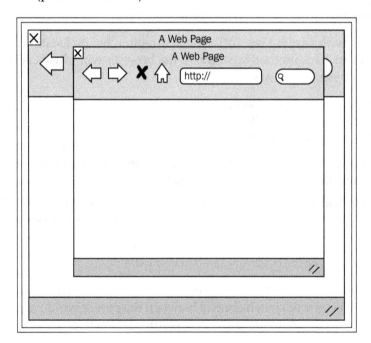

Multiple browser windows appears when:

- The user intends to click on a link from the parent browser window (for example, to share via social network)
- There is a sudden pop-up with an advertisement browser window

- Handling tests via the Internet Explorer web browser

  In the following example, a user is trying to switch control from a parent to a child window, perform tasks, close the driver, and then switch control from the child to the parent window:

```
String parentwindow = driver.getWindowHandle();
Set<String> handles =  driver.getWindowHandles();
for(String childwindow  : handles)
{
  if(!childwindow.equals(parentwindow))
  {
    driver.switchTo().window(childwindow);
    <!--Perform the steps for child|sub window here-->
    driver.close(); //closing child window
    //cntrl to main|parent window
    driver.switchTo().window(parentwindow);
  }
}
```

Select functions

A Select function allows you select or deselect values from a drop-down box or a radio button. It includes a list of Selenium API methods to work with select boxes that contain the `<select>...</select>` tags. These functions interact with the UI comboboxes to select options.

```
Select select = new Select(driver.findElement(By.
locatorType("path")));
```

Some helpful snippets using `select` functions are given below:

- The `selectByIndex(index)` method selects an option using the `index` value. The following is the syntax for this function:

```
select.selectByIndex(index);
```

  Let's get into the bookstore and select a product from a drop-down list. Here, we select the products at the top of the options list using the `selectByTndex()` method:

```
driver.get("http://www.barnesandnoble.com/");
Select select = new Select(driver.findElement(By.id("quick-search-
1-category")));
select.selectByIndex(1);
select.selectByIndex(2);
```

- The `selectByValue(value)` method selects an option using `value` in the string format. The following is the syntax for this function:

```
select.selectByValue("value");
```

Let's see how we can select an option from the bookstore's category list using the text value:

```
driver.get("http://www.barnesandnoble.com/");
Select select = new Select(driver.findElement(By.id("quick-search-1-category")));
select.selectByValue("music");
```

- The `getFirstSelectedOption()` method fetches the first selected option or currently selected option from the list.

```
select.getFirstSelectedOption();
```

Even if there are five options already picked or selected, this method recognizes only the first selected option. In the following code snippet, it returns the currently selected value at option 6:

```
driver.get("http://www.barnesandnoble.com/");
Select select = new Select(driver.findElement(By.id("quick-search-1-category")));
select.selectByIndex(6);
WebElement FSO = select.getFirstSelectedOption();
System.out.println(select.getFirstSelectedOption().getText());
```

- The `selectByVisibleText(text)` method selects an option from the select tag using text visibility. The following is the syntax for this function:

```
select.selectByVisibleText("text");
```

This method fetches text values, which are highly case sensitive. Let's search for a text value, `Music`, in the category list:

```
driver.get("http://www.barnesandnoble.com/");
Select select = new Select(driver.findElement(By.id("quick-search-1-category")));
select.selectByVisibleText("Music");
```

- The `getAllSelectedOptions()` method returns all the selected options. The following is the syntax for this function:

```
select.getAllSelectedOptions();
```

This method finds all the selected options from the list and returns them.
Let's see all the selected categories in this bookstore from the products list in
the following snippet of code:

```
driver.get("http://www.barnesandnoble.com/");
Select select = new Select(driver.findElement(By.id("quick-search-
1-category")));
List<WebElement> selectedOptions = select.getAllSelectedOptions();
for(WebElement b : selectedOptions) {
  System.out.println(b.getText());
}
```

- The getOptions() method returns all data from the options list. The
  following is the syntax for this function:

```
select.getOptions();
```

The list values of all the categories are returned from the bookstore, as shown
in the following snippet:

```
driver.get("http://www.barnesandnoble.com/");
Select select = new Select(driver.findElement(By.id("quick-search-
1-category")));
for (WebElement b : select.getOptions()) {
  System.out.println(b.getText());
}
```

- The isMultiple() method verifies multiple selection support. The following
  is the syntax for this function:

```
select.isMultiple();
```

For instance, the radio buttons or checkboxes have enough reason to have
multiple select features; however, some of them might not. This method
identifies the multi-select status and quickly responds to the user with
valuable feedback. Let's get into a bookstore with a categories (products)
combobox that doesn't involve any multi-select features as described:

```
driver.get("http://www.barnesandnoble.com/");
Select select = new Select(driver.findElement(By.id("quick-search-
1-category")));
if(select.isMultiple()){
System.out.println("Support multiple select at a time");
}
else{
System.out.println("Doesn't support multiple select at a time");
}
```

- The `deselectAll()` method deselects all the selected options. The following is the syntax of this function:

```
select.deselectAll();
```

A multi-select box with the options already chosen can obviously be deselected using this select method. Let's see an example in which a web page contains the multi-select box that is chosen with the default option. Finally, the `deselectAll()` method deselects all the chosen options as follows:

```
driver.get("http://compendiumdev.co.uk/selenium/basic_html_form.
html");
Select select = new Select(driver.findElement(By.
name("multipleselect[]")));
select.deselectAll();
```

- The `deselectByIndex(index)` method deselects an option using the `index` value. The following is the syntax for this function:

```
select.deselectByIndex(index);
```

This function is the reverse of the `selectByIndex(index)` method. Let's see how to deselect an option from the multi-select box using an integer value with the following piece of code:

```
driver.get("http://compendiumdev.co.uk/selenium/basic_html_form.
html");
Select select = new Select(driver.findElement(By.
name("multipleselect[]")));
select.deselectByIndex(3);
```

- The `deselectByValue(value)` method removes the specific option using `value`. The following is the syntax of this function:

```
select.deselectByValue(value);
```

This function is the reverse of the `selectByValue(value)` method. Let's see how to deselect an option from the multi-select box using a `String` value with the following piece of code:

```
driver.get("http://compendiumdev.co.uk/selenium/basic_html_form.
html");
Select select = new Select(driver.findElement(By.
name("multipleselect[]")));
select.deselectByValue("ms4");
```

- The `deselectByVisibleText(text)` method removes options using text visibility. The following is the syntax of this function:

```
select.deselectByVisibleText(text);
```

This function is the reverse of the `selectByVisibleText(text)` method. Let's see how to deselect an option from the multi-select box using `String` text with the following piece of code:

```
driver.get("http://compendiumdev.co.uk/selenium/basic_html_form.
html");
Select select = new Select(driver.findElement(By.
name("multipleselect[]")));
select.deselectByVisibleText("Selection Item 4");
```

Handling alerts and pop-ups

A pop-up is a browser window that opens randomly on surfing the Internet through the web browser. Web applications generate three different types of pop-ups, namely:

- JavaScript alert (pop-ups) (for example, advertisements)
- Browser pop-up (for example, a confirmation dialog box, an authentication prompt, and so on)
- Native OS pop-ups (for example, Windows pop-ups such as upload/download notfications)

JavaScript pop-ups are generally in the form of alerts and advertisements, especially for marketing purposes. Selenium WebDriver provides an API to handle the JavaScript pop-ups. The following is an example of an alert:

```
Alert alert = driver.switchTo().alert();
```

Some of the helpful snippets using JavaScript alert functions are as follows:

- The `dismiss()` function ignores or cancels the alert dialog box. The following is the syntax for this function:
  ```
  alert.dismiss();
  ```

  Let's see an example to wait for the alert dialog box, which is expected to appear, and then skip it to move forward:
  ```
  WebDriverWait wait = new WebDriverWait(driver, 10);
  try {
    wait.until(ExpectedConditions.alertIsPresent());
    Alert alert = driver.switchTo().alert();
    alert.dismiss();
  } catch (Exception e) {
    System.out.println("Alert is not available");
  }
  ```

- The `accept()` function acknowledges the alert dialog box upon our clicking the **OK** button.

```
alert.accept();
```

In general, the alert dialog box waits for an action to be initiated; it can be either alert cancellation or acceptance. Acceptance is one of two ways to close the alert pop-up, but with a reason. Let's see how to accept an alert dialog box by clicking on the **OK** button:

```
WebDriverWait wait = new WebDriverWait(driver, 10);
try {
  wait.until(ExpectedConditions.alertIsPresent());
  Alert alert = driver.switchTo().alert();
  alert.accept();
} catch (Exception e) {
  System.out.println("Alert is not available");
}
```

- The `getText()` function retrieves text or string values from the alert dialog box. The following is the syntax for this function:

```
alert.getText();
```

This method is normally used to verify the alert dialog box by fetching text values from the pop-up.

```
Alert alert = driver.switchTo().alert();
alert.getText();
```

- The `sendKeys()` method passes text or string values to the alert dialog box. The following is the syntax for this function:

```
alert.sendKeys(String arg0);
```

Sending text values into an alert pop-up dialog box is quite possible using these `alert` methods. Let's see a small piece of code to achieve this:

```
Alert alert = driver.switchTo().alert();
alert.sendKeys("Text to be passed");
```

- The `authenticateUsing()` method handles the basic HTTP authentication by passing the username and password to the browser pop-up. The following is the syntax for this function:

```
UserAndPassword user = new UserAndPassword("USERNAME",
"PASSWORD");
alert.authenticateUsing(user);
```

Authentication is a consistent blocker for many automation testers. Eventually, this method gets you into an environment to test with an API that helps you skip basic authentication blockers. Let's see an example to operate authentication pop-ups by passing valid credentials, as follows:

```
WebDriverWait wait = new WebDriverWait(driver, 10);
Alert alert = wait.until(ExpectedConditions.alertIsPresent());
alert.authenticateUsing(new UserAndPassword("USERNAME",
"PASSWORD"));
Actions action = new Actions(driver);
```

Mouse and keyboard actions

In Selenium WebDriver, actions can be either mouse actions or keyboard actions. The selenium-based Actions API provides support to click on a particular location with or without elements and use keyboard shortcuts efficiently.

- The build() method generates all the composite actions. The following is the syntax for this function:

```
action.build();
```

   The following code snippet explains to you how you can generate actions using the build() method:

```
Actions action = new Actions(driver);
action.click(driver.findElement(By.locatorType("path")));
action.build();
```

- The click() method performs a mouse click on the current mouse pointer location. The following is the syntax for this function:

```
action.click();
action.click(driver.findElement(By.locatorType("path")));
```

   Unlike clicking on the mouse's current location, this function allows you to click on an element by locating it (using the Actions class). Please check the following snippet for both cases:

```
driver.get("http://www.google.com");
Actions action = new Actions(driver);
action.click().build().perform();
action.click(driver.findElement(By.id("gsri_ok0"))).build().
perform();
```

- The `clickAndHold()` method lets you click and hold the current mouse pointer location. The following is the syntax for this function:

```
action.clickAndHold();
action.clickAndHold(driver.findElement(By.locatorType("path")));
```

Here, the selected source element is on hold unless the pressed key is released using the `release()` method. The following is a piece of code to click and hold an element:

```
action.clickAndHold().build().perform();
action.clickAndHold(driver.findElement(By.locatorType("path"))).
build().perform();
```

- The `contextClick()` method allows you to pop up the contextual menu, in instances such as when the right-click mouse operation occurs. The following is the syntax for this function:

```
action.contextClick();
action.contextClick(driver.findElement(By.locatorType("path")));
```

To do a context-click, it is important to locate an element before activating this action. Let's get through a snippet to understand the working of the `contextClick()` method:

```
driver.get("http://docs.seleniumhq.org");
Actions action = new Actions(driver);
WebElement ele = driver.findElement(By.xpath("//div[@
id='mainContent']/p[1]/i"));
action.moveToElement(ele).contextClick().build().perform();
```

- The `doubleClick()` method double-clicks on the current mouse pointer location. It also allows you to double-click on an element by locating it. The following is the syntax for this method:

```
action.doubleClick();
action.doubleClick(driver.findElement(By.locatorType("path")));
```

To double-click on an element, link text, link, context, or HTML5-based elements, the `doubleClick()` actions method is the right candidate. The following is a code snippet that selects a context by locating an element:

```
driver.get("http://docs.seleniumhq.org");
Actions action = new Actions(driver);
WebElement ele = driver.findElement(By.xpath("//div[@
id='mainContent']/p[1]/i"));
action.doubleClick(ele).build().perform();
```

- The `dragAndDrop()` function allows the user to drag an item from the source element and drop it on the target element. The following is the syntax for this function:

```
action.dragAndDrop(WebElement source, WebElement target);
```

Obviously, this method allows you to click and hold an element from a draggable source location, drag the element, and finally drop it in a suitable target location. Let's see how we can perform this action in the following example:

```
Actions action = new Actions(driver);
WebElement source = driver.findElement(By.locatorType("path"));
WebElement target = driver.findElement(By.locatorType("path"));
action.dragAndDrop(source, target).build().perform();
```

- The `dragAndDropBy()` function allows the user to drag an item from the source element and drop it on the target location using x and y coordinates. The following is the syntax of this function:

```
action.dragAndDropBy(WebElement source, int xOffset, int yOffset);
```

Unlike the `dragAndDrop()` method, this function uses offset values as a target location to release elements. Let's see how we can perform this action in the following example:

```
Actions action = new Actions(driver);
WebElement source = driver.findElement(By.locatorType("path"));
action.dragAndDropBy(source, 456, 234).build().perform();
```

- The `keyDown()` method allows the user to press or hold a specific key without releasing it. The following is the syntax for this function:

```
action.keyDown(Keys theKey);
```

Let's see an example that selects multiple keys using the control button in the pressed state:

```
Actions action = new Actions(driver);
WebElement source = driver.findElement(By.locatorType("path"));
WebElement target = driver.findElement(By.locatorType("path"));
action.keyDown(Keys.CONTROL);
action.click(source);
action.click(target);
action.keyUp(Keys.CONTROL);
action.perform();
```

- The `keyUp()` method releases the key that is already in a pressed condition. The following is the syntax for this function:

```
action.keyUp(Keys theKey);
```

The following example illustrates how we can move to the bottom of the page view using the *Ctrl* + *End* key combination and finally release the *Ctrl* key:

```
Actions action = new Actions(driver);
action.keyDown(Keys.CONTROL).sendKeys(Keys.END).keyUp(Keys.
CONTROL).build().perform();
```

- The `moveByOffset()` method moves the mouse pointer to a specific location using the *x* and *y* coordinates:

```
action.moveByOffset(int xOffset, int yOffset);
```

By default, the mouse offset will be located in the top-left corner of the page, that is, `(0,0)`. Customizing the *x* and *y* coordinates moves the mouse pointer to the desired location, as follows:

```
Actions action = new Actions(driver);
action.moveByOffset(234, 345).build().perform();
```

- The `moveToElement()` method moves the mouse pointer to a specific element. Besides this, the mouse pointer can also be moved from a source element to the target location using *xy* coordinates. The following is the syntax for this function:

```
action.moveToElement(WebElement source);
action.moveToElement(WebElement source, int xOffset, int yOffset);
```

This method directly scrolls the page view by highlighting the mouse pointer to the middle of the given element. Let's see how this method works with the following example:

```
Actions action = new Actions(driver);
WebElement source = driver.findElement(By.locatorType("path"));
action.moveToElement(source).build().perform();
action.moveToElement(source, 234, 345).build().perform();
```

- The `perform()` method executes actions. The following is the syntax for this function:

```
action.perform();
```

Though the `build()` function generates actions, the `perform()` method is the key to executing action commands. The following is a piece of code to generate and execute actions:

```
Actions action = new Actions(driver);
WebElement source = driver.findElement(By.locatorType("path"));
action.click(source);
action.build();
action.perform();
```

- The `release()` method drops an item fetched by the action of left-clicking the mouse. This function also releases an item to a specific element by locating it. The following is the syntax for this function:

```
action.release();
action.release(WebElement target);
```

Let's see how the `release()` method works with the drag and drop functionality in the following code snippet:

```
WebElement source = driver.findElement(By.locatorType("path"));
WebElement target = driver.findElement(By.locatorType("path"));
Action dragAndDrop = action.clickAndHold(source)
.moveToElement(target)
.release(target)
.build();
dragAndDrop.perform();
```

- The `sendKeys()` method controls the keyboard functions using keys. This function can also perform actions by locating elements. The following is the syntax for this function:

```
action.sendKeys(Keys theKeys);
action.sendKeys(Keys theKeys, Keys theKeys);
action.sendKeys(WebElement target, Keys theKeys);
```

This method supports Selenium users in three different flavors. Multiple combinations of keyboard shortcuts can be easily achieved using the `sendKeys()` method, as shown in the following code snippet:

```
Actions action = new Actions(driver);
WebElement target = driver.findElement(By.locatorType("path"));
action.sendKeys(Keys.TAB).build().perform();
action.sendKeys(Keys.CONTROL, Keys.END).build().perform();
action.sendKeys(target, Keys.TAB).build().perform();
```

Some of the helpful tasks using the above methods are given as follows:

- ° **Zoom In**: Use HTML tags for the advanced zoom-in and zoom-out features. The zoom-in feature gives better focus on a page under test. The following snippet explains well the zoom-in action on both Mac and Windows platforms:

```
WebElement html = driver.findElement(By.tagName("html"));
// WINDOWS
html.sendKeys(Keys.chord(Keys.CONTROL, Keys.ADD));
// MAC
html.sendKeys(Keys.chord(Keys.COMMAND, Keys.ADD));
```

- ° **Zoom Out**: The zoom out function is carried out by hitting the *Ctrl+-* key combinations. The following snippet explains well the zoom-out action:

```
WebElement html = driver.findElement(By.tagName("html"));
// WINDOWS
html.sendKeys(Keys.chord(Keys.CONTROL, Keys.SUBTRACT));
// MAC
html.sendKeys(Keys.chord(Keys.COMMAND, Keys.SUBTRACT));
```

- ° **Zoom 100%**: This snippet ensures that the page is fully loaded with 100 percent zoom capacity:

```
WebElement html = driver.findElement(By.tagName("html"));
// WINDOWS
html.sendKeys(Keys.chord(Keys.CONTROL, "0"));
// MAC
html.sendKeys(Keys.chord(Keys.COMMAND, Keys."0"));
```

- ° **Enter**: To submit a form or press the *Enter* key, the following piece of code will be helpful:

```
driver.findElement(By.locatorType("path")).sendKeys(Keys.
RETURN);
driver.findElement(By.locatorType("path")).sendKeys(Keys.
ENTER);
```

- ° **Drag and Drop**: Consider an application containing the drag and drop feature, in which an element is being dragged from one location to another. The source and target elements ought to be declared first so that the actions library prepares the drag and drop actions. Let's see the types of methods that allow you to perform a drag and drop functionality throughout the web page UI, as follows:

Method 1: In this model, the `dragAndDrop` action method is directly used to perform a drag and drop action by moving an item from the source element to the expected target element location:

```
WebElement source = driver.findElement(By.
locatorType("path"));
WebElement target = driver.findElement(By.
locatorType("path"));

Actions action = new Actions(driver);
Action dragAndDrop = action.dragAndDrop(source, target).
build();
dragAndDrop.perform();
```

Method 2: Here, the source and target elements are predefined, following a series of actions to click, hold, move, and release an element:

```
WebElement source = driver.findElement(By.
locatorType("path"));
WebElement target = driver.findElement(By.
locatorType("path"));
Actions action = new Actions(driver);
Action dragAndDrop = action.clickAndHold(source)
.moveToElement(target)
.release(target Element)
.build();
dragAndDrop.perform();
```

Method 3 (Using Java Robot for HTML5 pages): The Java `Robot` class is an optional method to drag and drop an element, similar to the Actions API. Let's see an example web page built with HTML5 technology that doesn't support Selenium's Actions method to drag and drop an element:

```
@Test
public void dragAndDrop() throws AWTException,
InterruptedException {

driver.get("http://demo.kaazing.com/forex/");
Actions action = new Actions(driver);
WebElement sourceElement = driver.findElement(By.
xpath("(//li[@name='dragSource'])[13]"));
Action drag = action.clickAndHold(sourceElement).build();
drag.perform();
```

```
WebElement targetElement = driver.findElement(By.
xpath("//section[@id='section1']/div[2]"));
Point coordinates = targetElement.getLocation();
Robot robot = new Robot(); //Robot for controlling mouse
actions
robot.mouseMove(coordinates.getX(), coordinates.getY() +
120);
Thread.sleep(2000);
robot.mouseMove(coordinates.getX(), coordinates.getY() +
110);
Thread.sleep(5000);
}
```

    °   **Mouse Hover**: When a user tries to hover over elements in a web page (for example, link text), the mouse hover events are automatically triggered to perform an action. Let's look at a snippet to mouse hover an element with the link text locator:

```
Actions action = new Actions(driver);
WebElement HoverLink = driver.findElement(By.
linkText("value"));
action.moveToElement(HoverLink);
action.perform();
```

 Refer to the following link to check all the available Actions keyboard keys: `https://sites.google.com/site/seleniumworks/keyboard-actions`

Summary

In this chapter, we learned about almost every single Selenium WebDriver function, along with their examples in detail.

In the next chapter, we will explore how we can practice Selenium WebDriver and learn about its techniques and how to survive without Selenium WebDriver functions.

4
Selenium WebDriver
Best Practices

In this chapter, we will learn the best practices for Selenium WebDriver and its techniques for handling a complex web application. Better understanding of WebDriver provides better results to find a quick solution. For example, JavascriptExecutor provides a quick workaround to automate web pages at a faster rate without using DOM.

In general, we face problems while involving an automation process without examining the application. It's quite necessary to understand why standard WebDriver approaches fail to work. Most of the problems occur when a page is overloaded with Ajax calls that load DOM elements asynchronously or when a page contains lots of frames, ActiveX/flex/flash components, and so on. Eventually, Selenium WebDriver overrides all these glitches with a proper workaround, where the driver simulates browsers exactly like a real user would do.

An efficient approach will help you to ensure better interaction with user interface components such as alerts, forms, and lists. For instance, the approach should be stable, quick, and reliable. The PageObject pattern is one of the best Selenium practices to maintain test suites or a collection of tests. Let's go through them one by one in the following sections.

In this chapter, we will learn the following topics:

- Handling Ajax websites
    - isElementPresent()
    - Waits
- Page Object pattern
- Event-firing WebDriver

- Handling iFrames
- Handling native OS and browser pop-ups
- JavascriptExecutor

Handling Ajax websites

Every modern web application makes use of Ajax calls that return data through asynchronous calls made to the web server. It avoids page reload and updates part of the web page at any time. Let's see how to manage these Ajax-based websites through Selenium WebDriver in detail.

The isElementPresent method

The isElementPresent() method is a user-defined method that checks for an element's availability within a web page. By default, the Selenium IDE generates the following script, where the object returns a Boolean value (however, this method is not recommended for handling Ajax-based web apps):

```
private boolean isElementPresent(By by) {
  try {
    driver.findElement(by);
    return true;
  } catch (NoSuchElementException e) {
    return false;
  }
}

@Test
public void Test01() throws Exception {
  driver.get("https://www.google.co.in/");
  Boolean a = isElementPresent(By.name("q"));
  System.out.println(a);
  Boolean b = isElementPresent(By.name("selenium_essentials"));
  System.out.println(b);
}
```

This method lets you wait for an element to execute appropriate actions. If the element is not found, it returns false. The try-catch statement in this example captures all the exceptions thrown (such as NoSuchElementException). This method can be written in many different ways. Let's see a few of the helpful tasks using this method:

- **Method 1**: This approach is used to verify whether an element is present in a page or not:

```
if (isElementPresent(By.locatorType("path"))) {
  System.out.println("Element is available");
} else {
  System.out.println("Element not available");
}
```

- **Method 2**: The following is an alternative method to check the element's availability in a web page:

```
if(driver.findElements(By.locatorType("path"))).size()>0)
{
  System.out.println("Element is present in the webpage");
} else {
  System.out.println("Element not available");
}
```

- **Method 3**: The following is a negative approach that verifies the unavailability of an element in a web page:

```
if(!isElementPresent(By.locatorType("path")))
{
  System.out.println("Element not available");
} else {
  System.out.println("Element is available");
}
```

In spite of the method verifying the element's availability, our tests certainly fail due to regular timeout (by returning `false`); what it lacks is `WebDriverWait` an explicit wait to meet a specific condition to occur and an implicit wait to wait for a specific time interval. We will explore this in detail in the following section.

Waits

Wait commands let you put tests on hold or pause for a few seconds or even days. Nowadays, Ajax-based websites are widely in used for their high data-exchanging speed. However, there will be variations in receiving each and every Ajax web service on a fully loaded page. To avoid such delays and to ignore exceptions such as `ElementNotVisibleException`, it is highly recommended to use waits. To handle such delays on a web page, Selenium WebDriver makes use of both implicit and explicit waits.

Explicit wait

An explicit wait waits for certain conditions to occur. It results in `TimeoutError` only when the conditions fail to meet their target. In general, the usage of the explicit wait is highly recommended. Here's an example of an explicit wait:

```
WebDriverWait wait = new WebDriverWait(driver, 10);
wait.until(ExpectedConditions.presenceOfElementLocated(By.
locatorType("path")));
```

In the example that follows, WebDriver waits for 20 seconds until the web element is found; if it is not, it will simply throw `TimeoutException`:

```
@Test
public void Test01() throws Exception {
   driver.get(baseUrl + "/");
try {
   waits().until(ExpectedConditions.presenceOfElementLocated
     By.id("invalidID")));
} catch(Exception e){
   System.out.println("Element Not Found");
}

private WebDriverWait waits(){
   return new WebDriverWait(driver, 20);
}
```

A few of the useful tasks that you can perform using explicit waits are listed as follows:

- **Handling an explicit wait on the PageObject pattern**: A `NullPointerException` exception will be thrown whenever `WebDriverWait` is declared globally on the PageObject design pattern. To avoid such risks, make use of the following snippet efficiently:

  ```
  public class classname{
    private WebDriver driver;
    private final Wait<WebDriver> wait;

    public classname(WebDriver driver) { //constructor
      this.driver = driver;
      wait = new WebDriverWait(driver, 20);
    }
  }

  public void Test01() throws Exception {
  ```

```
wait.until
(ExpectedConditions.presenceOfElementLocated(By.
locatorType("path")));
}
```

- **Handling an explicit wait by locators (ID, name, XPath, CSS, and so on)**:
  The following is an alternative method to utilize an explicit wait. It waits for
  a specific locator until the expected conditions are fulfilled. The locator can
  be an ID, name, XPath, or CSS, among others:

```
@Test
public void Test01() throws Exception {
  driver.get(baseUrl + "/");
  waitForID("value");
}

public void waitForID(String id) {
  WebDriverWait wait = new WebDriverWait(driver, 10);
  wait.until
(ExpectedConditions.presenceOfElementLocated(By.id(id)));
}
```

In the preceding code, `WaitForID` is a user-defined method, where the
locator `ID` can be customized with any locator types. `ExpectedCondition`
is one of the Selenium libraries containing a set of conditions that verify an
element's availability through `WebDriverWait`. The methods listed in the
following screenshot includes all the expected conditions of Selenium:

```
alertIsPresent()                              ExpectedCondition<Alert>
titleIs(String title)                         ExpectedCondition<Boolean>
elementSelectionStateToBe(By locat…           ExpectedCondition<Boolean>
elementSelectionStateToBe(WebEleme…           ExpectedCondition<Boolean>
elementToBeClickable(By locator)              ExpectedCondition<WebElement>
elementToBeClickable(WebElement…              ExpectedCondition<WebElement>
elementToBeSelected(By locator)               ExpectedCondition<Boolean>
elementToBeSelected(WebElement ele…           ExpectedCondition<Boolean>
frameToBeAvailableAndSwitchToIt(…             ExpectedCondition<WebDriver>
frameToBeAvailableAndSwitchToIt(…             ExpectedCondition<WebDriver>
invisibilityOfElementLocated(By lo…           ExpectedCondition<Boolean>
invisibilityOfElementWithText(By l…           ExpectedCondition<Boolean>
not(ExpectedCondition<?> condition)           ExpectedCondition<Boolean>
presenceOfAllElementsLocatedBy       ExpectedCondition<List<WebElem…
presenceOfElementLocated(By loc…    ExpectedCondition<WebElement>
refreshed(ExpectedCondition<T> condition)   ExpectedCondition<T>
stalenessOf(WebElement element)               ExpectedCondition<Boolean>
textToBePresentInElement(WebElemen…           ExpectedCondition<Boolean>
textToBePresentInElementLocated(By…           ExpectedCondition<Boolean>
textToBePresentInElement(By locato…           ExpectedCondition<Boolean>
textToBePresentInElementValue(By l…           ExpectedCondition<Boolean>
textToBePresentInElementValue(WebE…           ExpectedCondition<Boolean>
titleContains(String title)                   ExpectedCondition<Boolean>
visibilityOf(WebElement element)    ExpectedCondition<WebElement>
visibilityOfAllElements(L…    ExpectedCondition<List<WebElement>>
visibilityOfAllElementsLocatedBy    ExpectedCondition<List<WebEl…
visibilityOfElementLocated(By l…    ExpectedCondition<WebElement>
```

For further information, refer to the following link:
https://selenium.googlecode.com/git/docs/api/java/org/
openqa/selenium/support/ui/ExpectedConditions.html

The FluentWait method

The FluentWait method uses a polling technique, that is, it will keep on polling every fixed interval for a particular element to appear. The FluentWait method is more or less similar to explicit wait; however, FluentWait holds additional features, such as polling intervals and ignore exceptions. In the following example, the polling takes place every two seconds with a timeout of 10 seconds, so every two seconds, it checks for the element to appear, until it reaches 10 seconds, and the overall polling count will be five. Through FluentWait, the user can ignore any kind of exceptions, such as NoSuchElementException. The code for the FluentWait method is as follows:

```
@Test
public void Test01() throws Exception {
   driver.get(baseUrl + "/");
   fluentWait(By.locatorType("path"));
}

public WebElement fluentWait(final By locator) {
   FluentWait<WebDriver> wait = new FluentWait<WebDriver>(driver)
      .withTimeout(10, TimeUnit.SECONDS)
      .pollingEvery(2, TimeUnit.SECONDS)
      .ignoring(NoSuchElementException.class);

   WebElement foo = wait.until(new Function<WebDriver, WebElement>() {
      public WebElement apply(WebDriver driver) {
       return driver.findElement(locator);
      }
   });

   return foo;
};
```

Sleeper

The sleeper method is not an ideal approach to handle delays. The worst case of explicit wait is Thread.sleep, which is a fixed delay and mostly used for debugging test cases when the internet speed is slow. It lets the user pause test execution for a certain time period even though the expected condition is met.

The general usage of the sleeper method instead of the explicit wait or `fluentWait` is not appreciated. Here's the code for the sleeper method:

```
Thread.sleep(long millis);
```

Here, the sleeper is set to 3 seconds, that is, it lets you wait for 3 seconds to move forward to the next process:

```
Thread.sleep(3000); // waits for 3 secs
```

Timeouts

Unlike explicit wait, timeouts is an interface without any conditions that set a standard user-defined timeframe for a test to fail. The three types of timeouts are explained as follows:

- `implicitlyWait()`: An implicit wait waits for an element to appear or be displayed within a certain time period set by the user. In general, every timeout that occurs throughout the tests relies on the implicit wait. It lets you to poll the DOM for a specific time period until an element is found. An implicit wait acts as a master wait; however, the usage of the explicit wait or **Fluent Wait** is highly recommended since it has the ability to wait for dynamically loading Ajax elements that target a unique web element. The implicit wait is active from the start to the end of the test execution, that is, till the web page is closed. In the Selenium IDE, the wait defaults to 30 seconds. The following is the syntax for the implicit wait:

```
driver.manage().timeouts().implicitlyWait(long, TimeUnit);
```

  The following code snippet is one of the possible ways to define an implicit wait (it can be days, hours, microseconds, milliseconds, minutes, nanoseconds, or seconds):

```
driver.manage().timeouts().implicitlyWait(30, TimeUnit.DAYS);
driver.manage().timeouts().implicitlyWait(30, TimeUnit.HOURS);
driver.manage().timeouts().implicitlyWait(30, TimeUnit.
MICROSECONDS);
driver.manage().timeouts().implicitlyWait(30, TimeUnit.
MILLISECONDS);
driver.manage().timeouts().implicitlyWait(30, TimeUnit.MINUTES);
driver.manage().timeouts().implicitlyWait(30, TimeUnit.
NANOSECONDS);
driver.manage().timeouts().implicitlyWait(30, TimeUnit.SECONDS);
```

- `pageLoadTimeout()`: The `pageLoad` timeout method waits for the entire page to be loaded within a specific time period. A timeout exception will be thrown whenever a page takes more than the expected time to load. The following is the syntax for this method:

```
driver.manage().timeouts().pageLoadTimeout(long, TimeUnit);
```

In the example code snippet that follows, the `pageLoad` timer is set to 30 seconds for a web page to load soon after launching a browser under test:

```
driver.manage().timeouts().pageLoadTimeout(30, TimeUnit.SECONDS);
```

- `setScriptTimeout()`: The `setScript` timeout method waits for asynchronous APIs (Ajax) to be loaded in a web page within a certain time period. The following is the syntax for this method:

```
driver.manage().timeouts().setScriptTimeout(long, TimeUnit);
```

In the following example, the `setScript` timer is set to 30 seconds to avoid unusual Ajax timeout breaks. The timeout error returns whenever Ajax calls fail to retrieve within the given time period. The example code snippet follows:

```
driver.manage().timeouts().setScriptTimeout(30, TimeUnit.SECONDS);
```

The PageObject pattern

PageObject is an approach widely used in testing to reduce code duplication and increase the reusability of code. It is a design pattern that defines a page using objects. Moreover, the page object provides easy maintenance of code, and the scripts can be read easily anytime by any user. It is a pattern representing an entire page or a portion of the page in an object-oriented behavior.

Let's discuss the PageObject design pattern with an example on the Google web page. To start with the example, let's create a class that emphasizes how to write PageObject methods for a page in detail (`GoogleSearchPage.java`). The `search()` and `assertTitle()` methods let you perform Google searches and to assert the page title on the results page. Here's how we create the aforementioned class for this example:

```
public class GoogleSearchPage {
  public WebDriver driver;
  private final Wait<WebDriver> wait;

  public GoogleSearchPage(WebDriver driver) {
    this.driver = driver;
    wait = new WebDriverWait(driver, 10);
    //driver.get("https://seleniumworks.com");
```

```
    }

    public void search() throws Exception{
        //  search google
        wait.until.
  (ExpectedConditions.presenceOfElementLocated(By.name("q")));
        driver.findElement(By.name("q")).sendKeys("Prashanth Sams");
        driver.findElement(By.name("q")).submit();
        System.out.println("Google Search - SUCCESS!!");
        Thread.sleep(4000);
    }

    public void assertTitle() throws Exception{
        //  assert google search
        Boolean b = driver.getTitle().contains("Prashanth Sams");
        System.out.println(b);
    }
}
```

Now, create a test class using the TestNG unit test framework (for example, `TC.java`). The following test methods perform three different tasks, namely, opening the Google URL, searching for a keyword, and finally asserting the page title on the Google results page:

```
public class TC {
    private WebDriver driver;
    public GoogleSearchPage Task;

    @BeforeTest
    public void setUp() throws Exception {
        System.out.println("Instantiating Chrome Driver...");
        driver = new ChromeDriver();
    }

    @Test
    public void Test01() throws Exception {
        URL url = new URL(driver);
        url.geturl();
    }

    @Test
    public void Test02() throws Exception {
        Task = new GoogleSearchPage(driver);
        Task.search();
    }
```

```
  @Test
  public void Test03() throws Exception {
    Task = new GoogleSearchPage(driver);
    Task.assertTitle();
  }


  @AfterTest
  public void tearDown() throws Exception {
    driver.quit();
  }
}
```

Here, Selenium WebDriver is initiated at the start, followed by a series of test methods; the page objects are initialized inside the test class as well. The constructor avoids unusual errors by defining a driver as shown in the following Java class (URL.java). However, this class is significantly used to load the Google web page in this example. Once done with the task, it returns control to another class that contains PageObject methods for the tests to continue further. The following code exemplifies the discussion in this paragraph:

```
public class URL{
  public WebDriver driver;
  private String baseUrl;
  private boolean acceptNextAlert = true;
  private StringBuffer verificationErrors = new StringBuffer();

  public URL(WebDriver driver) {
    this.driver = driver;
    driver.get("http://www.google.co.in");
  }

  public GoogleSearchPage geturl() {
    System.out.println("Opened URL successfully");
    return new GoogleSearchPage(driver);
  }
}
```

The following is the final test report for all the preceding test cases:

TC (1): 3 total, 3 passed		6.08 s
		Collapse | Expand
packs		6.08 s
TC		6.08 s
Test01	passed	1.47 s
Opened URL successfully		
Test02	passed	4.58 s
Google Search - SUCCESS!!		
Test03	passed	37 ms
true		

```
===============================================
Custom suite
Total tests run: 3, Failures: 0, Skips: 0
===============================================
```

The PageFactory class

The `PageFactory` class is a class file under WebDriver's support library that maintains the PageObject pattern. It has the ability to find and locate elements quickly. The `PageFactory` class uses the `initElements` method to instantiate the WebDriver instance of PageObject. The following is the syntax for the the `PageFactory` class:

```
PageFactory.initElements(WebDriver driver, classname.class)
```

`NullPointerException` will be thrown when a user fails to implement `PageFactory`. By default, the `name` and `id` locator types can be accessed directly without labeling locators. For example, refer to the following code snippet:

```
private WebElement lst-ib; // Here, 'lst-ib' is an id
private WebElement q; // Here, 'q' is a name

lst-ib.sendKeys("selenium essentials");
q.sendKeys("Prashanth Sams");
```

Let's discuss `PageFactory` in detail with an example on the Google web page. This example is similar to the previous one. However, we will get to know how to make use of the `PageFactory` class efficiently in order to maintain the PageObject pattern. Let's create a test class using the TestNG unit-test framework (for example, `GoogleTest.java`). Initialize WebDriver and create an object using the `PageFactory` class as shown in the following snippet:

```java
public class GoogleTest {
  private WebDriver driver;

 @BeforeTest
  public void setUp() throws Exception {
    driver = new FirefoxDriver();
  }

 @Test
  public void Test01() throws Exception{
    GoogleSearchPage page = PageFactory.initElements(driver,
GoogleSearchPage.class);
    page.searchFor("Prashanth Sams");
  }

 @AfterTest
  public void Teardown() throws Exception{
    driver.quit();
  }
}
```

Now, create a class file (for example, `GoogleSearchPage.java`) containing methods where the locator values are declared without any attribute names:

```java
public class GoogleSearchPage {

  private WebElement q; // Here's the Element
  public WebDriver driver;

  public GoogleSearchPage(WebDriver driver) {
    this.driver = driver;
    driver.get("http://www.google.com/");
  }

  public void searchFor(String text) {
    q.sendKeys(text);
    q.submit();
  }
}
```

Here, the element `q` is declared without stating any locator type. However, it automatically identifies whether it is `id` or `name`. The PageFactory class finds an element with the `id` attribute matching; if not, it searches for the `name` attribute.

The @FindBy annotation

There are several methods to locate an element and one among them is the `@FindBy` annotation from PageFactory, which supports PageObject. The annotations `@FindBy` and `@FindBys` let you find elements using locators easily. The following snippets are an alternative approach to define the Google search text field. Here's the first of these snippets:

```
@FindBy(how = How.NAME, using = "q")
private WebElement searchBox;

@FindBy(name "q")
private WebElement searchBox;
```

Let's discuss this with an example using the `@FindBy` annotation on the Google web page. Here, the `@CacheLookup` annotation is used to keep the elements in cache and utilize them from the next time for rapid test execution:

```
import org.openqa.selenium.support.CacheLookup;
import org.openqa.selenium.support.FindBy;
import org.openqa.selenium.support.How;

public class GoogleSearchPage {

  @FindBy(how = How.NAME, using = "q") // or use @FindBy(name = "q")
  @CacheLookup
  private WebElement searchme;

  @FindBy(id = "lst-ib") // Less Verbose
  private WebElement submitme;
  //Here, both searchme & submitme refers to the same element

  public WebDriver driver;

  public GoogleSearchPage(WebDriver driver) {
    this.driver = driver;
    driver.get("http://www.google.com/");
  }

  public void searchFor(String text) {
```

```
        searchme.sendKeys(text);
        submitme.submit();
    }
}
```

The @FindBys annotation

The `@FindBys` annotation can have any numbers of tags in a series to define an element. In the following example, the `@FindBys` annotation handles three tags in a series to locate the Google search field. The tag series can be in top-to-bottom order, as shown in the following screenshot:

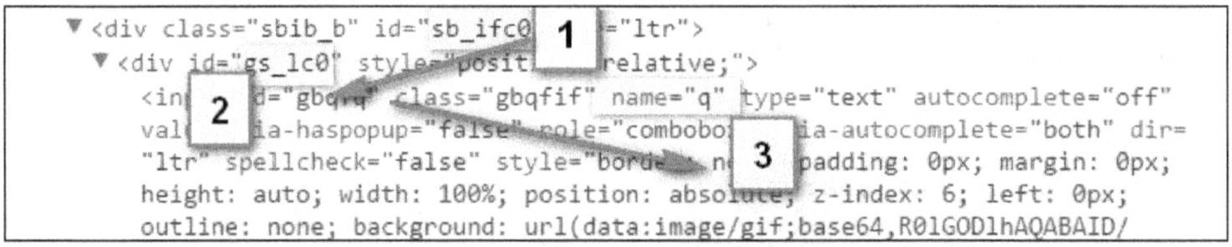

The following is the code for the `@FindBys` annotation:

```java
import org.openqa.selenium.support.CacheLookup;
import org.openqa.selenium.support.FindBy;
import org.openqa.selenium.support.FindBys;

public class GoogleSearchPage {

@FindBys({ @FindBy(id = "sb_ifc0"), @FindBy(id = "gs_lc0"), @
FindBy(name = "q") })
    @CacheLookup // keeps the element in cache for rapid execution
    private WebElement searchme;

    public WebDriver driver;

    public GoogleSearchPage(WebDriver driver) {
        this.driver = driver;
        driver.get("http://www.google.com/");
    }

    public void searchFor(String text) {
        searchme.sendKeys(text);
        searchme.submit();
    }
}
```

The EventFiringWebDriver class

The `EventFiringWebDriver` class is a sort of API that encloses arbitrary WebDriver instances to register `WebDriverEventListener`; it provides an automatic system to trigger events. For example, whenever a test failure occurs, the automatic screen capture method gets activated and starts logging screenshots. The `EventFiringWebDriver` class is highly recommended for better reporting and test maintenance. Through `EventFiringWebDriver`, the user can have full control over the web page under test, including control over navigation, over finding elements, on exception, the click action, over script execution, and so on.

Follow these regulations to run tests through `EventFiringWebDriver`:

1. Create a `Listeners` class by either implementing the `WebDriverEventListener` interface or extending the `AbstractWebDriverEventListener` class:

   ```
   public class Listeners extends AbstractWebDriverEventListener
   {...
   }


   public class Listeners implements WebDriverEventListener
   {...
   }
   ```

2. Now, create a test class and initialize `EventFiringWebDriver` (termed `eventDriver` in the following examples):

   ```
   private WebDriver driver;
   EventFiringWebDriver eventDriver = new
   EventFiringWebDriver(driver);
   ```

3. Finally, register the WebDriver event listener with `EventFiringWebDriver`:

   ```
   Listeners EL = new Listeners();
   eventDriver.register(EL);
   eventDriver.get("http://www.yourwebsite.com");
   ```

The user-defined WebDriver event listener functions are easily customizable and play a major role by responding to events. Let's discuss `EventFiringWebDriver` functions further, as follows:

- `afterNavigateBack()`: The `afterNavigateBack()` method is triggered soon after executing the `navigate().back()` command. The following is the syntax for this method:

  ```
  afterNavigateBack(WebDriver driver)
  ```

In the following code snippet, the `afterNavigateBack()` method prints the web page's current URL after navigating back from the previous URL:

```
@Override
public void afterNavigateBack(WebDriver driver) {
System.out.println("After Navigating Back. I'm at "
 + driver.getCurrentUrl());
}
```

- `afterNavigateForward()`: The `afterNavigateForward()` method is triggered soon after executing the `navigate().forward()` command. The following is the syntax for this method:

```
afterNavigateForward(WebDriver driver)
```

In the following code snippet, it prints the web page's current URL after navigating towards the previously visited URL:

```
@Override
public void afterNavigateForward(WebDriver driver) {
System.out.println("After Navigating Forward. I'm at "
 + driver.getCurrentUrl());
}
```

- `afterNavigateTo()`: The `afterNavigateTo()` method is triggered soon after executing the `get(String url)` or `navigate().to(String url)` commands. The following is the syntax for this method:

```
afterNavigateTo(java.lang.String url, WebDriver driver)
```

In the following code snippet, it prints the web page's current URL after visiting or opening a web page:

```
@Override
public void afterNavigateTo(String url, WebDriver driver) {
   System.out.println("After Navigating To: " + url + ", my url is:
"    + driver.getCurrentUrl());
}
```

- `beforeNavigateBack()`: The `beforeNavigateBack()` method is triggered just before executing the `navigate().back()` command. The following is the syntax for this method:

```
BeforeNavigateBack(WebDriver driver)
```

In the following code snippet, it prints the web page's current URL before navigating back from the previous URL:

```
@Override
public void beforeNavigateBack(WebDriver driver) {
  System.out.println("Before Navigating Back. I was at "
    + driver.getCurrentUrl());
}
```

- `beforeNavigateForward()`: The `beforeNavigateForward()` method is triggered just before executing the `navigate().forward()` command. The following is the syntax for this method:

  ```
  beforeNavigateForward(WebDriver driver)
  ```

  In the following code snippet, it prints the web page's current URL before navigating towards the previously visited URL:

  ```
  @Override
  public void beforeNavigateForward(WebDriver driver) {
    System.out.println("Before Navigating Forward. I was at "
      + driver.getCurrentUrl());
  }
  ```

- `beforeNavigateTo()`: The `beforeNavigateTo()` method is triggered just before executing the `get(String url)` or `navigate().to(String url)` commands. The following is the syntax for this method:

  ```
  beforeNavigateTo(java.lang.String url, WebDriver driver)
  ```

  In the following code snippet, it prints the web page's current URL before visiting or opening a web page:

  ```
  @Override
  public void beforeNavigateTo(String url, WebDriver driver) {
    System.out.println("Before Navigating To : " + url + ", my url
  was: "    + driver.getCurrentUrl());
  }
  ```

- `afterClickOn()`: The `afterClickOn()` method is triggered soon after executing the `WebElement.click()` command. The following is the syntax for this method:

  ```
  afterClickOn(WebElement element, WebDriver driver)
  ```

In the following code snippet, the `afterClickOn()` method prints the most recently clicked element:

```
@Override
public void afterClickOn(WebElement element, WebDriver driver) {
System.out.println("Clicked Element with: '" + element + "'");
}
```

- `beforeClickOn()`: The `beforeClickOn()` method is triggered just before executing `WebElement.click()` command. The following is the syntax for this method:

```
beforeClickOn(WebElement element, WebDriver driver);
```

In the following code snippet, the element likely to be clicked is highlighted just before executing the `click()` function:

```
@Override
public void beforeClickOn(WebElement element, WebDriver driver) {
   System.out.println("Trying to click: '" + element + "'");
   /** Highlight Elements before clicking **/
   for (int i = 0; i < 1; i++) {
      JavascriptExecutor js = (JavascriptExecutor) driver;
      js.executeScript("arguments[0].setAttribute('style',
arguments[1]);", element, "color: black; border: 3px solid
black;");
   }
}
```

- `afterFindBy()`: The `afterFindBy()` method is triggered soon after executing the `WebElement.findElement()`, `WebElement.findElements()`, `driver.findElement()`, or `driver.findElements()` command. The following is the syntax for this method:

```
afterFindBy(By by, WebElement element, WebDriver driver);
```

In the following code snippet, `afterFindBy()` prints the element soon after locating it:

```
private By finalFindBy;
@Override
public void afterFindBy(By by, WebElement element, WebDriver
driver) {
   finalFindBy = by;
   System.out.println("Found: '" + finalFindBy + "'.");
   // This is optional or an alternate method
   System.out.println("Found: " + by.toString() + "'.");
}
```

- `beforeFindBy()`: The `beforeFindBy()` method is triggered just before executing the `WebElement.findElement()`, `WebElement.findElements()`, `driver.findElement()`, or `driver.findElements()` command. The following is the syntax for this method:

  `beforeFindBy(By by, WebElement element, WebDriver driver);`

  In the following code snippet, the `beforeFindBy()` method prints the element just before locating it:

```
private By finalFindBy;
@Override
public void beforeFindBy(By by, WebElement element, WebDriver
driver) {
  finalFindBy = by;
  System.out.println("Trying to find: '" + finalFindBy + "'.");
  System.out.println("Trying to find: " + by.toString()); // This
is optional or an alternate method
}
```

- `afterScript()`: The `afterScript()` method is triggered soon after executing the `RemoteWebDriver.executeScript(java.lang.String, java.lang.Object[])` command. The following is the syntax of this method:

  `afterScript(java.lang.String script, WebDriver driver)`

  In the following code snippet, `afterScript()` prints a successful message soon after executing the JavaScript function:

```
@Override
public void afterScript(String script, WebDriver driver) {
  System.out.println("JavaScript is Executed");
}
```

- `beforeScript()`: The `beforeScript()` method is triggered just before executing the `RemoteWebDriver.executeScript(java.lang.String, java.lang.Object[])` command. The following is the syntax for this method:

  `beforeScript(java.lang.String script, WebDriver driver)`

  In the following code snippet, `beforeScript()` prints a foretelling message, saying that the JavaScript is about to be executed:

```
@Override
public void beforeScript(String script, WebDriver driver) {
  System.out.println("JavaScript is about to execute");
}
```

- `afterChangeValueOf()`: The `afterChangeValueOf()` method is triggered soon after executing the `WebElement.clear()` or `WebElement.sendKeys()` command. The following is the syntax of this method:

```
afterChangeValueOf(WebElement element, WebDriver driver);
```

In the following code snippet, this method prints the text field values available both before and after the Google search:

```
private String actualValue;
@Override
public void afterChangeValueOf(WebElement element, WebDriver
driver) {
  String modifiedValue = "";
  try {
    modifiedValue = element.getText();
  } catch (StaleElementReferenceException e) {
System.out.println("Could not log change of element, because of a
stale" + " element reference exception.");
    return;
  }

  // What if the element is not visible anymore?
  if (modifiedValue.isEmpty()) {
    modifiedValue = element.getAttribute("value");
  }

System.out.println("The value changed from '" + actualValue + "'
to '" + modifiedValue + "'");
}
```

- `beforeChangeValueOf()`: The `beforeChangeValueOf()` method is triggered just before executing the `WebElement.clear()` or `WebElement.sendKeys()` command. The following is the syntax of this method:

```
beforeChangeValueOf(WebElement element, WebDriver driver)
```

In the following code snippet, this method prints the value that exists before clearing the Google search field:

```
private String actualValue;
@Override
public void beforeChangeValueOf(WebElement element, WebDriver
driver) {
  actualValue = element.getText();

  // What if the element is not visible anymore?
```

```
    if (actualValue.isEmpty()) {
      actualValue = element.getAttribute("value");
    }

    System.out.println("The existing value is: " + actualValue);
    }
```

- onException(): The onException() method is triggered whenever an exception is thrown during test execution. The following is the syntax of this method:

  ```
  onException(java.lang.Throwable throwable, WebDriver driver)
  ```

  In the following code snippet, the EventFiringWebDriver takes a screenshot soon after getting an exception thrown on test execution:

  ```
  @Override
  public void onException(Throwable throwable, WebDriver webdriver)
  {

    File scrFile = ((TakesScreenshot) webdriver)
      .getScreenshotAs(OutputType.FILE);
    try {
      org.apache.commons.io.FileUtils.copyFile(scrFile, new
  File("C:\\Testfailure.jpeg"));
    } catch (Exception e) {
      System.out.println("Unable to Save");
    }
  }
  ```

Event-firing WebDriver example

Let's discuss this with an example by creating a Listener class, which is said to be an EventListener class. Here, we are extending the AbstractWebDriverEventListener class instead of implementing the WebDriverEventListener class; however, either class does the same job. Through EventListener, you can override any listeners as you wish and the events can be fired on any popular browsers; for example, the following class is named Listeners.java:

```
public class Listeners extends AbstractWebDriverEventListener {
  // public class Listeners implements WebDriverEventListener {
    private By finalFindBy;
    private String actualValue;
```

As discussed in the preceding functions with an exercise on each, there are different events that are triggered each time `EventFiringDriver` is executed. The following is a list of navigation functions with user-defined steps to perform an action:

```
/** URL NAVIGATION | navigate(), get() **/
// Prints the URL before Navigating to specific URL
@Override
public void beforeNavigateTo(String url, WebDriver driver) {
System.out.println("Before Navigating To : " + url + ", my url was: "
+ driver.getCurrentUrl());
}

// Prints the current URL after Navigating to specific URL
@Override
public void afterNavigateTo(String url, WebDriver driver) {
System.out.println("After Navigating To: " + url + ", my url is: " +
driver.getCurrentUrl());
}

// Prints the URL before Navigating back "navigate().back()"
@Override
public void beforeNavigateBack(WebDriver driver) {
System.out.println("Before Navigating Back. I was at " + driver.
getCurrentUrl());
}

// Prints the current URL after Navigating back from the previous URL
@Override
public void afterNavigateBack(WebDriver driver) {
System.out.println("After Navigating Back. I'm at " + driver.
getCurrentUrl());
}

// Prints the URL before Navigating forward "navigate().forward()"
@Override
public void beforeNavigateForward(WebDriver driver) {
System.out.println("Before Navigating Forward. I was at " + driver.
getCurrentUrl());
}

// Prints the current URL after Navigating forward "navigate().
forward()"
@Override
public void afterNavigateForward(WebDriver driver) {
System.out.println("After Navigating Forward. I'm at " + driver.
getCurrentUrl());
}
```

The following methods are automatically activated on locating an element. They help you to identify whether an element is available in a page or not:

```
/** FINDING ELEMENTS | findElement(), findElements() **/
// Called before finding Element(s)
 @Override
public void beforeFindBy(By by, WebElement element, WebDriver driver)
{
  finalFindBy = by;
  System.out.println("Trying to find: '" + finalFindBy + "'.");
  System.out.println("Trying to find: " + by.toString()); // This is
optional and an alternate way
}

 // Called after finding Element(s)
 @Override
public void afterFindBy(By by, WebElement element, WebDriver driver) {
  finalFindBy = by;
  System.out.println("Found: '" + finalFindBy + "'.");
  System.out.println("Found: " + by.toString() + "'."); // This is
optional and an alternate way
 }
```

The following are used to debug and are also helpful during an ongoing presentation. Every testing scenario involves clicking on a link and verifying the expected page; EventFiringWebDriver does more than an ordinary click() function would do, as shown in the following code:

```
/** CLICK | click() **/
// Called before clicking an Element
@Override
public void beforeClickOn(WebElement element, WebDriver driver) {
  System.out.println("Trying to click: '" + element + "'");

  // Highlight Elements before clicking
  for (int i = 0; i < 1; i++) {
    JavascriptExecutor js = (JavascriptExecutor) driver;
    js.executeScript("arguments[0].setAttribute('style',
arguments[1]);", element, "color: black; border: 3px solid black;");
  }
}

// Called after clicking an Element
@Override
public void afterClickOn(WebElement element, WebDriver driver) {
System.out.println("Clicked Element with: '" + element + "'");
}
```

The following methods maintain a clear tracking system by storing both existing and updated values:

```
/** CHANGING VALUES | clear(), sendKeys() **/
// Before modifying values
@Override
public void beforeChangeValueOf(WebElement element, WebDriver driver)
{
  actualValue = element.getText();

  if (actualValue.isEmpty()) {
    actualValue = element.getAttribute("value");
  }
System.out.println("The existing value is: " + actualValue);
}


// After modifying values
@Override
public void afterChangeValueOf(WebElement element, WebDriver driver) {
  String modifiedValue = "";
  try {
    modifiedValue = element.getText();
  } catch (StaleElementReferenceException e) {
    System.out.println("StaleElementReferenceException is thrown");
    return;
  }

  if (modifiedValue.isEmpty()) {
    modifiedValue = element.getAttribute("value");
  }

  System.out.println("The value changed from '" + actualValue + "' to
'" + modifiedValue + "'");
}
```

The following methods can operate before and after executing RemoteWebDriver, which is composed of both client and server. Shown here is an example with JavaScript execution:

```
/** JAVASCRIPT | beforeScript(), afterScript()**/
// Called before executing RemoteWebDriver.executeScript(java.lang.
String, java.lang.Object[])
@Override
public void beforeScript(String script, WebDriver driver) {
    System.out.println("JavaScript is about to execute");
```

```
}

// Called after executing RemoteWebDriver.executeScript(java.lang.
String, java.lang.Object[])
@Override
public void afterScript(String script, WebDriver driver) {
  System.out.println("JavaScript is Executed");
}
```

The following method, which allows the user to capture screens on test failure, is one of the most important methods used in `EventFiringWebDriver`:

```
/** ON EXCEPTION | capture screenshots on test failure **/
// Takes screenshot on any Exception thrown during test execution
@Override
public void onException(Throwable throwable, WebDriver webdriver) {
  System.out.println("Caught Exception");
  File scrFile = ((TakesScreenshot) webdriver)
    .getScreenshotAs(OutputType.FILE);
  try {
    org.apache.commons.io.FileUtils.copyFile(scrFile, new File("C:\\
Testfailure.jpeg"));
  } catch (Exception e) {
    System.out.println("Unable to Save");
  }
}
```

Let's define a `Test` class to run events through the `Listeners` class. To do so, initialize `EventFiringWebDriver` along with `EventListenerDriver`, and finally register `Listeners` to the `EvenFiringWebDriver` instance. The following test class explicates all the `EventFiringWebDriver` functions on a Google web page in detail:

```
private WebDriver driver;
@BeforeTest
public void setUp() throws Exception {
  driver = new FirefoxDriver();
}

@Test
public void Test01() throws Exception {

  EventFiringWebDriver eventDriver = new EventFiringWebDriver(driver);
  Listeners EL = new Listeners();
  eventDriver.register(EL);
```

```
   // beforeNavigateTo | afterNavigateTo
   eventDriver.get("http://www.bing.com");

   // beforeNavigateBack | afterNavigateBack
   eventDriver.get("http://www.google.com");
   eventDriver.navigate().back();

   // beforeNavigateForward | afterNavigateForward
   eventDriver.navigate().forward();

   // beforeFindBy | afterFindBy
   eventDriver.findElement(By.name("q"));

   // beforeClickOn | afterClickOn
   eventDriver.findElement(By.id("sblsbb")).click();

   // afterScript() | beforeScript()
   JavascriptExecutor jse = (JavascriptExecutor) eventDriver;
   jse.executeScript("alert('Selenium Essentials saved my Day!')");

   // beforeChangeValueOf | afterChangeValueOf
   eventDriver.findElement(By.name("q")).sendKeys("Selenium
Essentials");

   // onException
   eventDriver.findElement(By.id("Wrong Value"));
}

@AfterTest
public void tearDown() throws Exception {
   driver.quit();
}
```

Handling iframes

A web page can have any number of iframes (inline frames) to represent new pages inside a main page. They can be either multiple iframes or nested iframes. The iframes are indicated with an iframe tag, such as <iframe>...</iframe>.

It's easy to handle iframes when a user discovers all the iframes available in a web page. Google Chrome's Developer debugging tool is used to check the availability of iframes. The following figure is an example of nested iframes:

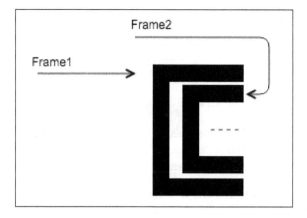

To handle iframes, it's important to switch into and move out of an iframe to the main frame. The following is the syntax for switching iframes:

```
driver.switchTo().frame()
```

The following code snippet is a real-time example of switching iframes to locate elements.

```
driver.switchTo().frame(driver.findElement(By.locatorType("iframe[id=
'Value']")));
```

To access an iframe located outside the present iframe, it is essential to terminate the current iframe. Closing an iframe lets you move from the current iframe to the main content. To do so, follow the following syntax:

```
driver.switchTo().defaultContent(); // close iFrame
```

To locate an element in a web page with nested iframes, try the `switchTo()` method multiple times to switch between iframes. The following code snippet is a sample nested iframe structure, followed by a snippet for handling multiple iframes:

```
<iframe ...>
  <iframe ...>
    <iframe ...>
    </iframe>
  </iframe>
</iframe>
```

The following is a snippet that handles the structured nested iframes:

```
driver.switchTo().frame(driver.findElement(By.locatorType("iframe[
id='1']")));
driver.switchTo().frame(driver.findElement(By.locatorType("iframe[
id='2']")));
driver.switchTo().frame(driver.findElement(By.locatorType("iframe[
id='3']")));
driver.switchTo().defaultContent();
driver.switchTo().defaultContent();
driver.switchTo().defaultContent();
```

Obtain the iframe position to handle iframes without id or name. To do so, use the following syntax:

```
driver.switchTo().frame(value);
```

The following iframes are available at the first and second positions:

```
driver.switchTo().frame(0);
driver.switchTo().frame(1);
```

Handling native OS and browser pop-ups using Java Robot

Some of the most popular UI-based automation tools, such as AutoIT, Sikuli, and Java Robot, are quite easy to integrate with Selenium WebDriver tests. However, it is difficult to implement and operate these tools on varying screen resolutions, handling Selenium Grid, cross-browser tests, and more. In general, the Selenium WebDriver API doesn't support native OS and browser pop-up handling. The browser profile is a set of customized browser instances, which is an alternative choice to handle these pop-ups.

The Java.awt.Robot library is a Java library file that supports Selenium WebDriver to interact with web applications through control over mouse and keyboard actions.

Let's discuss the following example to perform a simple Google search using Java Robot and Selenium WebDriver:

```
import java.awt.Robot;
import java.awt.event.KeyEvent;
import java.lang.reflect.Field;

public class classname {
  private WebDriver driver;
  private String baseUrl;
```

```
@Test
public void Test01() throws Exception {
    driver = new FirefoxDriver();
    driver.get("https://www.google.com");

    Robot r = new Robot();
    driver.findElement(By.name("q")).click();
    Thread.sleep(4000);
    typeKeys("Prashanth Sams", r);
}

public static void typeKeys(String str, Robot r) {
  for (int i = 0; i < str.length(); i++) {
    typeCharacter(r, "" + str.charAt(i));
  }
}

public static void typeCharacter(Robot robot, String letter) {
  try {
  boolean upperCase = Character.isUpperCase(letter.charAt(0));
  String variableName = "VK_" + letter.toUpperCase();
  Class c = KeyEvent.class;
  Field field = c.getField(variableName);
  int keyCode = field.getInt(null);
  robot.delay(1000);

  if (upperCase)
    robot.keyPress(KeyEvent.VK_SHIFT);
    robot.keyPress(keyCode);
    robot.keyRelease(keyCode);

  if (upperCase)
    robot.keyRelease(KeyEvent.VK_SHIFT);
  } catch (Exception e) {
    System.out.println(e);
  }
}
```

Downloading browser pop-ups

Multiplatform support, such as running tests through Selenium Grid with varying screen resolutions, is not feasible on Java Robot, and it always depends on the screen (x,y) coordinates. In general, the browser pop-up is in the form of a download dialog box, upload dialog box, advertisements, and more.

Let's see how to save a file from a browser download pop-up as shown in the following screenshot:

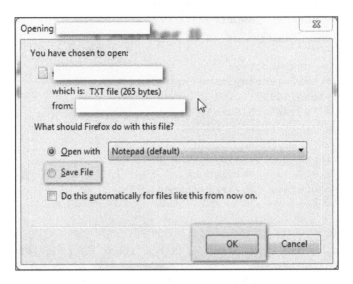

The prerequisites for the next code snippet to execute are:

* Native window screen resolution of **1920 x 1080**
* FF Browser window status of **Maximize**
* Customized *xy* coordinates according to the screen resolution

Make sure that all the preceding specifications are met before executing the following piece of code. Here, it is mandatory to customize location coordinates for screens with different resolutions. In this example, the Java `Robot` class is intended to choose **Save file** on clicking the radio button and finally clicking on the **OK** button:

```
Robot r = new Robot();
/** click Save File **/
r.mouseMove(787, 544); //move to co-ordinate Location
r.mousePress(InputEvent.BUTTON1_MASK); //Left Mouse click-Press
r.mouseRelease(InputEvent.BUTTON1_MASK); //Left Mouse click-Release
r.delay(5); //wait for 5 millisecs
/** click ok **/
r.mouseMove(10322, 641); //move to co-ordinate Location
r.mousePress(InputEvent.BUTTON1_MASK); //Left Mouse click-Press
r.mouseRelease(InputEvent.BUTTON1_MASK); //Left Mouse click-Release
```

Screen capture

Screen capture provides explicit test reports by logging test failures as screenshots. Furthermore, the Java `Robot` class is also helpful in taking instant screenshots on each test failure. The following is the code for this method:

```
Robot r = new Robot();
BufferedImageimg = r.createScreenCapture(new Rectangle(0, 0, 100,
100));
File path = new File("C://screen.jpg");
ImageIO.write(img, "JPG", path);
```

There are several ways to capture the screen. Let's see some of the helpful methods, which are as follows:

- **Method 1**: Selenium provides the augmenter to take screenshots in any given timeframe of test execution:

```
WebDriver augmentedDriver = new Augmenter().augment(driver);
File screenshot = ((TakesScreenshot)augmentedDriver).
getScreenshotAs(OutputType.FILE);
String path = "/Users/prashanth_sams/Desktop/" + screenshot.
getName();
FileUtils.copyFile(screenshot, new File(path));
```

- **Method 2**: Selenium provides another method to capture screens using Java `Robot` as follows:

```
java.awt.Dimension size = Toolkit.getDefaultToolkit().
getScreenSize();
Robot r = new Robot();
BufferedImage img = r.createScreenCapture(new Rectangle(size));
File path = new File("C://screen.jpg");
ImageIO.write(img, "JPG", path);
```

In the following list are some of the most important `Robot` class functions that are eventually valuable while integrating Selenium WebDriver:

```
Robot r = new Robot();
r.mouseMove(500, 500); //move to co-ordinate Location
r.mousePress(InputEvent.BUTTON1_MASK); //Left Mouse click-Press
r.mouseRelease(InputEvent.BUTTON1_MASK); //Left Mouse click-Release
r.mousePress(InputEvent.BUTTON2_MASK); //Middle Mouse click-Press
r.mouseRelease(InputEvent.BUTTON2_MASK); //Middle Mouse click-Release
r.mousePress(InputEvent.BUTTON3_MASK); //Right Mouse click-Press
r.mouseRelease(InputEvent.BUTTON3_MASK); // Right Mouse click-Release
r.mouseWheel(7);  //Scroll Mouse
r.getPixelColor(500, 100); //Get Pixel color-RBG
```

```
MouseInfo.getPointerInfo().getLocation(); //Get Current Mouse Location
Toolkit.getDefaultToolkit().getScreenSize(); //Get Screen Resolution-
Dimension
r.createScreenCapture(new Rectangle(size)); //Screen capture
r.keyPress(KeyEvent.VK_ENTER); //Press Enter Key
r.keyRelease(KeyEvent.VK_ENTER); //Release Enter Key
r.delay(5); //wait for certain milliseconds
```

Refer to the following link to check all the Java Robot keyboard actions:
`https://sites.google.com/site/seleniumworks/java_robot`
Mofiki's Coordinate Finder lets you find the instant screen coordinates of
the current mouse pointer location.

Firefox profile to download files

Whenever a user tries to download a file, they get a download dialog box that keeps
on asking whether the file has to be saved or opened with an application. The simplest
way to ignore these browser pop-ups and to save files is through browser profiles.
It can be done either manually or through setting up Firefox profile preferences.

The following couple of methods let you download files locally without any risk:

- **Method 1**: Here is a quick workaround to download files through manually
  setting up the Firefox profile's default behavior. This method certainly
  provides a manual alternative to the next method, which uses set preferences
  to disable the Firefox browser's download pop-up. Follow these steps to
  make this quick difference:

  1. Launch the Firefox web browser.

  2. Go to the Firefox applications under the **Tools** menu
     (**Tools** | **Options** | **Applications**).

3. Replace/set all the download actions to **Save File**, as shown in the following screenshot:

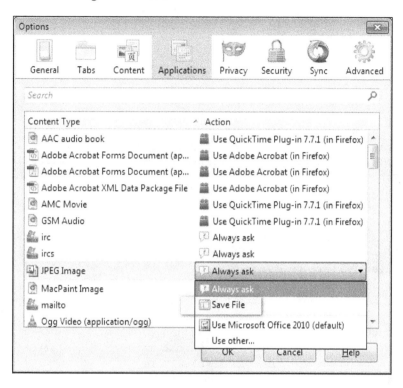

4. Click on the **OK** button and restart the browser.

- **Method 2**: By customizing the Firefox profile through setting preferences, we can directly download files without any external disturbances. The code snippet is summarized in the following steps:

  1. Initialize the Firefox profile as follows:

     ```
     FirefoxProfile profile = new FirefoxProfile();
     ```

  2. Set your preference for all the file types that prompt to save, as follows:

     ```
     "browser.helperApps.neverAsk.saveToDisk"
     ```

Refer to the following code to avoid the Firefox browser's default download pop-up through `setPreferences`:

```
FirefoxProfile profile = new FirefoxProfile();

String path = "C:\\Test\\";
profile.setPreference("browser.download.folderList", 2);
profile.setPreference("browser.download.dir", path);
profile.setPreference("browser.download.manager.alertOnEXEOpen",
false);
profile.setPreference("browser.helperApps.neverAsk.saveToDisk",
"application/msword, application/csv, application/ris, text/csv,
image/png, application/pdf, text/html, text/plain, application/zip,
application/x-zip, application/x-zip-compressed, application/download,
application/octet-stream");
profile.setPreference("browser.download.manager.showWhenStarting",
false);
profile.setPreference("browser.download.manager.focusWhenStarting",
false);
profile.setPreference("browser.download.useDownloadDir", true);
profile.setPreference("browser.helperApps.alwaysAsk.force", false);
profile.setPreference("browser.download.manager.alertOnEXEOpen",
false);
profile.setPreference("browser.download.manager.closeWhenDone", true);
profile.setPreference("browser.download.manager.showAlertOnComplete",
false);
profile.setPreference("browser.download.manager.useWindow", false);
profile.setPreference("services.sync.prefs.sync.browser.download.
manager.showWhenStarting", false);
profile.setPreference("pdfjs.disabled", true);

driver = new FirefoxDriver(profile);
```

The JavascriptExecutor class

`JavascriptExecutor` is a class under the Selenium library that executes JavaScript code snippets. For example, a Selenium WebDriver command such as `WebElement.click()` might not work on all browsers, but `JavaScriptExecutor` could help you to click on an element in any browser by executing the appropriate JavaScript snippet. There are a couple of Selenium WebDriver functions to handle JavaScript, such as `executeAsyncScript()` and `executeScript()`. The `executeAsyncScript()` method lets you inject JavaScript snippets into the page for asynchronous execution. The following is the syntax for this method:

```
JavascriptExecutor jse = (JavascriptExecutor)driver;
```

Let's see how you can click on an element without using Selenium's `click()` method. The following code snippet displays an alert message as a pop-up:

```
(JavascriptExecutor) driver).executeScript("arguments[0].click();",
WebElement element);
JavascriptExecutor jse = (JavascriptExecutor)driver;
jse.executeScript("alert('Selenium Essentials saved my Day!')");
```

A few of the helpful tasks using JavascriptExecutor are discussed as follows.

Page scroll

Page scroll allows the user to focus on an element to perform any action. It helps you to scroll throughout the web page using *xy* coordinates. There are numerous methods and techniques to make it happen.

The following is a list of methods to scroll the web page:

- **Scroll down**: This is the `JavascriptExecutor` method that makes use of the *xy* coordinates to scroll down the page. The page scrolls down with respect to the changes made on the *y* co-ordinate. The following is the code for this method:

```
import org.openqa.selenium.JavascriptExecutor;
JavascriptExecutor jse = (JavascriptExecutor)driver;
jse.executeScript("scroll(0, 250)"); //Y-coordinate '250' can be
altered
```

- **Scroll up**: Similar to the above snippet, JavaScript uses *xy* coordinates to scroll up the page. The page scrolls up with respect to the changes made on the *x* coordinate. The following is the code for this method:

```
JavascriptExecutor jse = (JavascriptExecutor)driver;
jse.executeScript("scroll(250, 0)"); // X-coordinate '250' can be
altered
```

- **Scroll to bottom of the page**: The `document.documentElement.scrollHeight` method returns the height of the HTML element. The `document.body.scrollHeight` method outputs the page/frame height, and finally `document.documentElement.clientHeight` returns the browser window height. This method gets the user to focus on the bottom of the page quickly. The following is the code for this method:

```
JavascriptExecutor jse = (JavascriptExecutor)driver;
jse.executeScript("window.scrollTo(0,Math.max(document.
documentElement.scrollHeight,document.body.scrollHeight,document.
documentElement.clientHeight));");
```

- **Full scroll to bottom in slow motion**: The following two methods get the user focus on the bottom of the page in slow motion.
  - ° **Method 1:**
    ```
    for(int second = 0;; second++) {
      if (second >= 60) {
        break;
      }
      ((JavascriptExecutor) driver).executeScript("window.
    scrollBy(0,400)", ""); //y value '400' can be altered
        Thread.sleep(3000);
    }
    ```

  - ° **Method 2:**
    ```
    JavascriptExecutor jse = (JavascriptExecutor) driver;
    for (int second = 0;; second++) {
      if (second >= 60) {
        break;
      }
    }
    jse.executeScript("window.scrollBy(0,800)", ""); //y value
    '800' can be altered
    Thread.sleep(3000);
    }
    ```

- **Scroll automatically to WebElement**: These methods automatically focus on a web page element by using different types of JavaScript functions:
  - ° **Method 1:**
    ```
    Point hoverItem =driver.findElement(By.xpath("Value")).
    getLocation();
    ((JavascriptExecutor)driver).executeScript("return window.
    title;");
    Thread.sleep(6000);
    ((JavascriptExecutor)driver).executeScript("window.
    scrollBy(0,"+(hoverItem.getY())+");");
    // Adjust your page view by making changes right over here
    (hoverItem.getY()-400)
    ```

  - ° **Method 2:**
    ```
    ((JavascriptExecutor)driver).executeScript("arguments[0].
    scrollIntoView();", driver.findElement(By.
    xpath("Value')]")));
    ```

○ **Method 3**:

```
WebElement element = driver.findElement(By.
xpath("Value"));
Coordinates coordinate = ((Locatable)element).
getCoordinates();
coordinate.onPage();
coordinate.inViewPort();
```

Highlighting elements

Debugging can be even easier by highlighting elements at runtime through the browser UI. Optimizing user-interactive CSS values paves the way for better user interactions while conducting a seminar or a demonstration. A similar function is available in the Selenium IDE as an add-on, **Highlight Elements (Selenium IDE)**, for stress-free debugging. The following is the code for this function:

```
WebElement element1 = driver.findElement(By.className("Value"));
WebElement element2 = driver.findElement(By.id("Value"));

JavascriptExecutor jse = (JavascriptExecutor)driver;
jse.executeScript("arguments[0].setAttribute('style', arguments[1]);",
element1, "color: blue; border: 2px solid blue;");
jse.executeScript("arguments[0].setAttribute('style', arguments[1]);",
element2, "color: yellow; border: 0px solid red;");
```

The following code snippet provides you with a better understanding of customizing CSS values through JavaScript:

```
@Test
public void Highlight() throws Exception {
  driver.get(baseUrl + "/");
  WebElement searchbutton = driver.findElement(By.id("Value"));
  highlightElement(searchbutton);
  WebElement submitbutton = driver.findElement(By.id("Value"));
  highlightElement(submitbutton);
  element.click();
}

public void highlightElement(WebElement element) {
  for (int i = 0; i < 2; i++) {
    JavascriptExecutor js = (JavascriptExecutor) driver;
    js.executeScript("arguments[0].setAttribute('style',
arguments[1]);", element, "color: yellow; border: 2px solid yellow;");
    js.executeScript("arguments[0].setAttribute('style',
arguments[1]);", element, "");
  }
}
```

Opening a new browser window

Tests on a single browser instance are quite usual and consistent. However, opening a new browser window at the start or in the middle of the test lets you open a new browser instance as a child window. So, you have two windows open at a time with an active, newly opened browser window.

The following snippet lets you open a new browser instance to make it ready for testing:

```
@Test
public void Test01() throws Exception {
  OpenNewTab("https://www.google.com");
}
public void trigger(String script, WebElement element) {
  ((JavascriptExecutor) driver).executeScript(script, element);
}
public Object trigger(String script) {
  return ((JavascriptExecutor) driver).executeScript(script);
}

public void OpenNewTab(String url) {
  String script = "var d=document,a=d.createElement('a');a.target='_
blank';a.href='%s';a.innerHTML='.';d.body.appendChild(a);return a";
  Object element = trigger(String.format(script, url));
  if (element instanceof WebElement) {
    WebElement anchor = (WebElement) element;
    anchor.click();
    trigger("var a=arguments[0];a.parentNode.removeChild(a);",
anchor);
  } else {
    throw new JavaScriptException(element, "Unable to open Window",
1);
  }
}
```

In the preceding code snippet, OpenNewTab is the user-defined method that opens a new browser window along with the given URL.

JavaScript error collector

The JSErrorCollector library is a third-party, external Java library file that collects all the JavaScript errors through the Firefox profile during test execution. This library file feeds an extension, JSErrorCollector.xpi, into the Firefox browser profile on runtime to collect JS errors. All the captured JS errors are listed as counts in the bottom-right corner of the browser's page. The following screenshot depicts a list of captured JS errors:

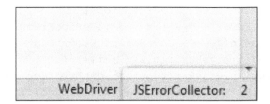

Follow these steps to store the JavaScript errors generated at runtime:

1. Download the `JSErrorCollector` Java library file from the following link:

    ```
    http://cl.ly/3W160N1R0E0U/JSErrorCollector-0.5.jar
    ```

2. Add the downloaded JAR file to build the path.

3. Initialize the Firefox profile with the `JSErrorCollector` extension,
    as follows:

    ```
    FirefoxProfile ffProfile = new FirefoxProfile();
    JavaScriptError.addExtension(ffProfile);
    ```

4. Finally, collect all the JavaScript errors:

    ```
    List < JavaScriptError > jsError = JavaScriptError.
    readErrors(driver);
    ```

The following code lets you collect the JS errors and reflects them in the console
after test execution:

```
import java.util.List;
import net.jsourcerer.webdriver.jserrorcollector.JavaScriptError;
import org.openqa.selenium.firefox.FirefoxProfile;

@BeforeTest
public void setUp() throws Exception {
  FirefoxProfile ffProfile = new FirefoxProfile();
  JavaScriptError.addExtension(ffProfile);
  driver = new FirefoxDriver(ffProfile);
  baseUrl = "http://404checker.com";
}

@Test
public void Test01() throws Exception {
  driver.get(baseUrl + "/");
  Thread.sleep(5000);
}

@AfterTest
```

```
public void tearDown() throws Exception {
List < JavaScriptError > jsError = JavaScriptError.readErrors(driver);
   System.out.println("————JavaScript Error List————");
   for (int i = 0; i < jsError.size(); i++) {
     System.out.println(jsError.get(i).getErrorMessage());
System.out.println("Error Line: " + jsError.get(i).getLineNumber());
     System.out.println(jsError.get(i).getSourceName());
   }
   System.out.println("————End of the List————");
   System.out.println("\n");
   driver.close();
   driver.quit();
}
```

The following screenshot is a test result for the preceding script that displays the error type, the error line, and the affected JavaScript filename.

```
————JavaScript Error List————
TypeError: a is null
Error Line: 17
http://pagead2.googlesyndication.com/pagead/js/graphics.js
————End of the List————
```

Summary

In this chapter, you learned the best practices for Selenium WebDriver on how to make use of WebDriver techniques and how to survive through external libraries when WebDriver support is not in use.

In the next chapter, we will discuss Selenium frameworks and their unique approach to build successful projects.

5
Selenium WebDriver Frameworks

A test automation framework helps reduce the repetition of the same task again and again. In general, the automation framework increases productivity by raising code reusability and reducing coding efforts and test maintenance. Selenium WebDriver is exceptionally robust in building test automation frameworks. Every framework starts with a prototype version followed by stabilized standard versions. The Selenium WebDriver framework consists of three significant flavors, namely, Data-Driven, Keyword-Driven, and Hybrid-Driven frameworks.

Once the framework has been developed, its structure can be accessed by different projects and is reusable. Thus, it avoids building a new Selenium test automation framework from scratch. The key role of this chapter is to explain how to create a test automation framework for use in Selenium projects.

In this chapter, we will learn the following topics:

- Behavior-Driven Development
    - Cucumber BDD framework
    - JBehave BDD framework

- JXL API Data-Driven framework
    - Read and write Excel sheet
    - Simple Data-Driven approach
    - Selenium Data-Driven testing using reusable library
    - Data-Driven testing using TestNG | `@dataProvider`

- Apache POI Data-Driven framework
    - ° Read and write Binary Workbook (`.xls`)
    - ° HSSF usermodel (`.xls`)
    - ° XSSF usermodel (`.xlsx`)
    - ° SS usermodel (`.xls`, `.xlsx`]

- Text file Data-Driven framework
- Properties file Data-Driven framework
- CSV file Data-Driven framework
- Keyword-Driven framework
- Hybrid-Driven framework

Behaviour-Driven Development

Behaviour-Driven Development (BDD) is a part of the successful Agile methodology that creates a mutual understanding between business analysts, testers, and developers. It is a practice extended from the **Test-Driven Development** (TDD) and **Acceptance-Driven Development (ATDD)** approaches. Acceptance tests enclose business logic with procedural steps and tasks that fully rely on the software requirements, that is, they determine whether or not the given test scenarios meet with the requirements. The BDD scenarios are easy to understand and are reusable, especially their maintenance which is effortless in any time period. They can be parameterized using tables as a data source; on the other hand, multiple scenarios let you handle end-to-end tests as well. Refer to the following scenario:

```
Scenario: User authentication

Given I am at the login page
When I enter valid credentials
Then I should log in
```

Gherkin-based frameworks such as Cucumber and JBehave are fairly attractive and robust. Gherkin is a human (business) readable Domain Specific Language (DSL).

Cucumber BDD framework

Cucumber is a tool developed in the Ruby language that automates acceptance test scenarios by matching stories with step definitions using patterns. Here, the stories are documented as plain text and featured in a `.feature` file format. Cucumber can be implemented and is available for most of the popular languages, such as Java, .Net, Ruby, and so on. See the following for a sample `Feature` file format:

Feature: Log in to an e-commerce site

Scenario: Real user authentication

Given I am at the login page
When I enter valid credentials
Then I should log in

Cucumber JVM

Cucumber JVM is a Java implementation of the Cucumber tool that can be integrated with Selenium WebDriver to run web-based automation tests. It is recommended that all the Eclipse IDE users add the `cucumber-eclipse` plugin through **Help | Install new software...**, with the input URL, `http://cucumber.github.com/cucumber-eclipse/update-site`.

Maven is the safest build automation tool to execute Cucumber JVM tests. First, create a Maven project and add the plugin along with the following dependencies in the generated `POM.xml` file:

```xml
<build>
  <plugins>
    <plugin>
      <groupId>org.apache.maven.plugins</groupId>
      <artifactId>maven-compiler-plugin</artifactId>
      <version>2.5.1</version>
      <configuration>
        <encoding>UTF-8</encoding>
        <source>1.6</source>
        <target>1.6</target>
      </configuration>
    </plugin>
    <plugin>
      <groupId>org.apache.maven.plugins</groupId>
```

```
        <artifactId>maven-surefire-plugin</artifactId>
        <version>2.12.2</version>
        <configuration>
          <useFile>false</useFile>
        </configuration>
      </plugin>
    </plugins>
  </build>
  <dependencies>
    <dependency>
      <groupId>junit</groupId>
      <artifactId>junit</artifactId>
      <version>4.11</version>
      <scope>test</scope>
    </dependency>
    <dependency>
      <groupId>org.testng</groupId>
      <artifactId>testng</artifactId>
      <version>6.8</version>
    </dependency>
    <dependency>
      <groupId>org.seleniumhq.selenium</groupId>
      <artifactId>selenium-server</artifactId>
      <version>2.43.1</version>
    </dependency>
    <dependency>
      <groupId>info.cukes</groupId>
      <artifactId>cucumber-core</artifactId>
      <version>1.1.2</version>
    </dependency>
    <dependency>
      <groupId>info.cukes</groupId>
      <artifactId>cucumber-java</artifactId>
      <version>1.1.2</version>
    </dependency>
    <dependency>
      <groupId>info.cukes</groupId>
      <artifactId>cucumber-junit</artifactId>
      <version>1.1.2</version>
    </dependency>
    <dependency>
      <groupId>info.cukes</groupId>
      <artifactId>cucumber-html</artifactId>
      <version>0.2.2</version>
```

```
      </dependency>
      <dependency>
        <groupId>info.cukes</groupId>
        <artifactId>gherkin</artifactId>
        <version>2.11.6</version>
      </dependency>
      <dependency>
        <groupId>org.hamcrest</groupId>
        <artifactId>hamcrest-core</artifactId>
        <version>1.3</version>
      </dependency>
      <dependency>
        <groupId>info.cukes</groupId>
        <artifactId>gherkin</artifactId>
        <version>2.11.6</version>
      </dependency>
      <dependency>
        <groupId>com.rubiconproject.oss</groupId>
        <artifactId>jchronic</artifactId>
        <version>0.2.6</version>
        <scope>test</scope>
      </dependency>
  </dependencies>
```

The following screenshot depicts a project structure for the given example:

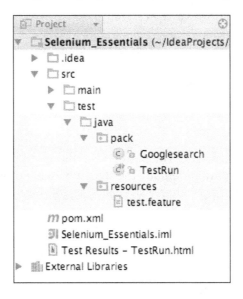

Let's see an example with Google search that asserts a search keyword on the Google results page using Cucumber JVM through Selenium WebDriver:

1. Start BDD automation tests by creating specs in a `.feature` extension file (`test.feature`). A single feature can have any number of scenarios. The following is the code for this step:

```
@foo
Feature: test


Scenario Outline: google search case#1


Given Google page "http://www.google.com"
When I enter the keyword "<keyword>" in search box
When I press enter key
Then I should get the results of "<expected>"


Examples: ASSERT GOOGLE SEARCH1
  | keyword             | expected            |
  | prashanth sams      | prashanth sams      |
  | selenium essentials | selenium essentials |
  | seleniumworks       | seleniumHQ          |


Scenario Outline: google search case#2


Given Google page "http://www.google.com"
When I enter the keyword "<keyword>" in search box
When I press enter key
Then I should get the results of "<expected>"


Examples: ASSERT GOOGLE SEARCH2
  | keyword             | expected            |
  | prashanth sams      | prashanth sams      |
  | selenium essentials | selenium essentials |
```

Cucumber defines a set of rules called regular expression patterns (regex patterns) to pass parameters for the steps written. Some of the most commonly used regex patterns are listed as follows:

Regex patterns	Description
[0-9]* or \d*	This is used to pass a series of digits.
[0-9]+ or \d+	This is used to pass a series of digits except an empty string.
\d+\.\d+	This is used to pass floating point numbers.

Regex patterns	Description
"[^\"]*"	This is used to pass anything in double quotes.
.	This is used to pass only one character.
.*	This is used to pass any number of characters.
.+	This is used to pass any character except an empty string.

2. Create step definitions using the @Given, @When, @And, @But, and @Then annotations provided by the Cucumber API (Googlesearch.java), as follows:

```java
public class Googlesearch {

    WebDriver driver;

    @Before
    public void beforeTest() {
        driver = new FirefoxDriver();
    }

    @Given("^Google page \"([^\"]*)\"$")
    public void I_open_google_page_as(String URL) throws Throwable {
        driver.get(URL);
        driver.manage().timeouts().implicitlyWait(10, TimeUnit.SECONDS);
    }

    @When("^I enter the keyword \"(.*)?\" in search box$")
    public void i_enter_in_search_box(String arg1) throws Throwable
    {
        driver.findElement(By.name("q")).sendKeys(arg1);
    }

    @When("^I press enter key$")
    public void i_press_enter_key() throws Throwable {
        driver.findElement(By.name("q")).submit();
    }

    @Then("^I should get the results of \"(.*)?\"$")
    public void i_should_see_results_of(String content) throws
    Throwable {
        Thread.sleep(3000L);
        Boolean b = driver.getPageSource().contains(content);
        Assert.assertTrue(b);
```

```
    }

    @After
    public void after(Scenario scenario) {
        driver.close();
        driver.quit();
      }
    }
```

Finally, create a global test class to execute Cucumber JVM tests (TestRun.java). Customize the feature file location and format as mentioned in the following class:

```
import org.junit.runner.RunWith;
import cucumber.api.junit.Cucumber;

@RunWith(Cucumber.class)@Cucumber.Options(format = {
"pretty", "html:target/cucumber", "json:target/cucumber.json"
}, features = "/Users/prashanth_sams/IdeaProjects/Selenium_
Essentials/src/test/java/resources")

public class TestRun {
}
```

Use the maven clean command to wipe out files and directories generated in the previous test run. The command, maven install lets you run Cucumber tests. The Cucumber JVM builds a clear test report through the Surefire Report plugin (as mentioned in the preceding example). Ant is also a build automation tool similar to Maven and therefore is an alternative for executing tests.

JBehave BDD framework

JBehave is an open source, Java-based BDD, whereas rbehave is an implementation for the Ruby language. It is an efficient framework for Behavior-Driven Development, similar to the Cucumber JVM. Here, the stories are documented as plain text and featured in the .story file format. Annotations such as @Given, @When, and @Then are used to bind the textual map steps to Java methods. The base stories have to be configured as per the requirements for running stories. To do so, you need a separate class for configuration (Basestories.java). JBehave has the embeddable interface to run stories, whereas the configuration instances configure the running of the stories and CandidateSteps instances match textual steps in the stories. These two embedders, namely configurable embedders and injectable embedders are used to run stories.

The JUnit-enabled embeddables are listed as follows:

- JUnitStory
- JUnitStories

In general, for one-to-one mapping, the superclass used is JUnitStory; for many-to-one mapping, the superclass used is JUnitStories. Find the two types of classes, one-to-one mapping and many-to-one mapping, at the JBehave official site at `http://jbehave.org/reference/stable/developing-stories.html`.

Let's see an example with Google search that asserts a search keyword on the Google results page using JBehave through Selenium WebDriver:

Start BDD automation tests by creating specifications in the `.story` file format (`googlesearch.story`), as shown in the following code:

```
Scenario:  Google search
Given I have a google welcome page
When I search for <keywords>
Then I should get the <expected> text in the results page
Examples:
 | keywords           | expected  |
 | prashanth sams     | prashanth |
 | selenium essentials | prashanth |
```

Then, create a new story class, `Basestories.java`, which is the heart of the entire story-running process. Check the following link to get the modified JBehave stories class for one-to-one mapping at `https://gist.github.com/prashanth-sams/642f8bc1ebcff76f98d2`.

One-to-one mapping is used in this example since we have only one story and a single-step class, but we are extending `JUnitStories`. Here, the `Basestories` class is an abstract class that requires no base class; you have a notable subclass for a configurable embedder, that is, the base class that extends `JUnitstories`. The `configuration()`, `stepsFactory()`, and `storyPaths()` methods are extended from the `JUnitStories` class. The annotation `override` is used here on extending the `JUnitStories` class. Generally, the configurable embedder allows subclasses such as base stories to specify the configuration steps and the injectable embedder allows the injection of the fully specified embedder. Here, the `createsteps()` method is used to run tests from another class, and the `storyPaths()` method locates the story file. All the Eclipse IDE users are recommended to add the `JBehave-eclipse` plugin from **Help | Install new software...** using the URL, `http://jbehave.org/reference/eclipse/updates/`.

The following screenshot depicts a project structure for the following example:

The steps class is clearly a Pojo. Pojo is short for a plain old Java object that does not extend any class and is literally called a Java class. It encloses the Java methods to be mapped with the aforementioned human-readable story file. Create a test step class using annotations provided by the JBehave API (`Googlesteps.java`):

```java
public class Googlesteps {

    private int result;
    private WebDriver driver;
    private String baseUrl;

    @Given("I have a google welcome page")
    public void GoogleWelcomePage() {
        driver = new ChromeDriver();
        driver.manage().timeouts().implicitlyWait(30, TimeUnit.SECONDS);
        driver.get("https://www.google.co.in/");

    }

    @When("I search for <keywords>")
    public void searchGoogle(@Named("keywords") String keywords) throws
InterruptedException {
```

```
    driver.findElement(By.name("q")).clear();
    driver.findElement(By.name("q")).sendKeys(keywords);
    driver.findElement(By.name("sblsbb")).click();
    Thread.sleep(2000);
  }


@Then("I should get the <expected> text in the results page")
  public void searchResult(@Named("expected") String expected) {
    String bodyText = driver.findElement(By.tagName("body")).
getText();
    Assert.assertTrue("Selenium", bodyText.contains(expected));
    driver.close();
    driver.quit();
  }


}
```

In the preceding example, `Googlestories.java` is the test class used to execute the stories. The base class, `Basestories.java`, is extended here since it is an abstract class and one cannot create objects in it:

```
public class Googlestories extends Basestories {

  @Override
  protected List < Object > createSteps() {
    List < Object > steps = new ArrayList < Object > ();
    steps.add(new jbehave.google.steps.Googlesteps());
    // TODO Auto-generated method stub
    return steps;
  }

  @Override
  public void run() throws Throwable {
    // TODO Auto-generated method stub
    super.run();
  }
}
```

The stories can be executed through build automation tools such as Ant and Maven or using development IDEs such as Eclipse, IntelliJ IDEA, and so on. The following screenshot displays the test result obtained with one pass condition and one fail condition:

```
Running story jbehave/google/stories/googlesearch.story
Starting ChromeDriver (v2.9.248307) on port 46532
Starting ChromeDriver (v2.9.248307) on port 34524

(jbehave/google/stories/googlesearch.story)
Scenario: Google search
Examples:
Given I have a google welcome page
When I search for <keywords>
Then I should get the <expected> text in the results page

|keywords|expected|
|prashanth sams|prashanth|
|selenium essentials|prashanth|

Example: {keywords=prashanth sams, expected=prashanth}
Given I have a google welcome page
When I search for prashanth sams
Then I should get the prashanth text in the results page

Example: {keywords=selenium essentials, expected=prashanth}
Given I have a google welcome page
When I search for selenium essentials
Then I should get the prashanth text in the results page (FAILED)
(java.lang.AssertionError: Selenium)
```

Multiple scenarios can be executed in the JBehave BDD framework. Use $ to identify the parameters; however, it can also be customized using the base class explained in the previous section. Refer to the following code snippets to run multiple JBehave acceptance test scenarios:

```
Scenario:  Google search #1
Given I have a google welcome page
When I search for seleniumessentials
Then I should get the selenium text in the results page

Scenario:  Google search #2
Given I have a google welcome page
When I search for seleniumworks.com
Then I should get the selenium text in the results page
```

The following is a test class for the preceding multiple scenarios:

```
@Given("I have a google welcome page")
public void GoogleWelcomePage() {
   driver = new ChromeDriver();
   driver.manage().timeouts().implicitlyWait(30, TimeUnit.SECONDS);
```

```
  driver.get("https://www.google.co.in/");
}

@When("I search for $keyword")
public void searchGoogle(String keyword) throws InterruptedException {
  driver.findElement(By.name("q")).sendKeys(keyword);
  driver.findElement(By.id("sblsbb")).click();
}

@Then("I should get the $expected text in the results page")
public void searchResult(String expected) {
  String bodyText = driver.findElement(By.tagName("body")).getText();
  Assert.assertTrue("Selenium", bodyText.contains(expected));
}
```

JXL API Data-Driven framework

Data-Driven testing is a software-testing methodology, which is an iterative process to assert the actual value with the expected value that fetches test input data from an external or internal data source. Data-Driven tests are generally carried out with bulk input data.

Java Excel API is termed the **JXL API**. It is the most widely used API for executing Selenium Data-Driven tests that allows the user to read, write, create, and modify sheets in an Excel Binary (.xls) workbook at runtime. JXL API has no support for SpreadsheetML (.xlsx) workbooks.

Reading and writing in an Excel sheet

The API reads data from Excel Binary workbooks with versions Excel 95, 97, 2000, XP, and 2003. The following snippet tells you how to read an Excel Binary workbook:

```
FileInputStream fi = new FileInputStream("C:\\...\inputdata.xls");
Workbook wb = Workbook.getWorkbook(fi);
Sheet ws = wb.getSheet(0);
String a[][] = new String[ws.getRows()][ws.getColumns()];

for (int rowCnt = 1; rowCnt < ws.getRows(); rowCnt++) {
  driver.get("www.example.com");
  driver.findElement(By.locatorType("path")).sendKeys(ws.getCell(0,
rowCnt).getContents());
  ...
}
```

Here, the `getWorkbook()` method fetches the workbook as a file and not as string; the `getSheet()` method accesses the sheet of the workbook; and the `getRows()` and `getColumns()` methods store row and column counts using two-dimensional arrays.

The JXL library also allows the users to create an Excel workbook and write data into the workbook. The following snippet tells you how to write an Excel Binary workbook:

```
FileOutputStream fo = new FileOutputStream("C:\\...\outputdata.xls");
WritableWorkbook wb = Workbook.createWorkbook(fo);
WritableSheet ws = wb.createSheet("Sheet1", 0);

for (int rowCnt = 1; rowCnt < wrksheet.getRows(); rowCnt++) {
  driver.get("www.example.com");
  driver.findElement(By.locatorType("path")).sendKeys(wrksheet.
getCell(0, rowCnt).getContents());
  driver.findElement(By.locatorType("path")).click();

  boolean resultfound = isElementPresent(By.locatorType("path"));

  if (resultfound) {
    //Writes data into 3rd column
    Label l3 = new Label(2, rowCnt, "pass");
    ws.addCell(l3);
    else {
      //Writes data into 3rd column
      Label l2 = new Label(2, rowCnt, "fail");
      ws.addCell(l2);
    }
  }
}
wb.write();
wb.close();
}
```

Here, the `write()` method returns the values to be saved in the Excel workbook and the `close()` method quits the current workbook session. The `getContents()` method returns all the values from the cell. Here's the syntax for this method:

```
wrksheet.getCell(0, rowCnt).getContents()
```

Let's see an example test method to **read and write data in an Excel Binary** workbook:

```
@Test
public void readandwrite() throws Exception {

  // Read data from excel sheet
  FileInputStream fi = new FileInputStream("C:\\...\inputdata.xls");
```

```
Workbook wrkbook = Workbook.getWorkbook(fi);
Sheet wrksheet = wrkbook.getSheet(0);
String a[][] = new String[wrksheet.getRows()][wrksheet.
getColumns()];
// Write the input data into another excel file
FileOutputStream fo = new FileOutputStream("C:\\...\outputdata.
xls");
WritableWorkbook wwb = Workbook.createWorkbook(fo);
WritableSheet ws = wwb.createSheet("customsheet", 0);

System.out.println("Total Rows: " + wrksheet.getRows());
System.out.println("Total Columns: " + wrksheet.getColumns());

for (int i = 0; i < wrksheet.getRows(); i++) {

    for (int j = 0; j < wrksheet.getColumns(); j++) {
        a[i][j] = wrksheet.getCell(j, i).getContents();
        Label l = new Label(j, i, a[i][j]);
        Label l1 = new Label(2, 0, "Result");
        ws.addCell(l);
        ws.addCell(l1);
    }
}

for (int rowCnt = 1; rowCnt < wrksheet.getRows(); rowCnt++) {

    driver.get("www.example.com");
  //Enter search keyword by reading data from Excel [Here it read
from 1st column]

    driver.findElement(By.locatorType("path")).sendKeys(wrksheet.
getCell(0, rowCnt).getContents());
    driver.findElement(By.locatorType("path")).click();
    Thread.sleep(5000);

    boolean resultfound = isElementPresent(By.locatorType("path"));

    if (resultfound) {
      //Writes data into 3rd column
      Label l3 = new Label(2, rowCnt, "pass");
      ws.addCell(l3);
      else {
        //Writes data into 3rd column
        Label l2 = new Label(2, rowCnt, "fail");
        ws.addCell(l2);
```

```
        }
      }
    wwb.write();
    wwb.close();
}
```

Simple Data-Driven approach

The Data-Driven technique is a repetitive process of any step with multiple sets of data. The Selenium WebDriver API doesn't have any built-in support for Data-Driven tests. Regardless of that, the JExcel/JXL library is a third-party API for Selenium-based tests to perform Data-Driven tasks. In this approach, the script to activate the Data-Driven function is embedded in the test class itself instead of the reusable library being put to use. Let's see an example with a pass condition and a fail condition using a simple Data-Driven technique from JXL API. The following screenshot is an Excel data source of the given example:

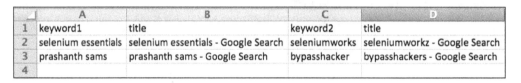

	A	B	C	D
1	keyword1	title	keyword2	title
2	selenium essentials	selenium essentials - Google Search	seleniumworks	seleniumworkz - Google Search
3	prashanth sams	prashanth sams - Google Search	bypasshacker	bypasshackers - Google Search
4				

Here, we search for a set of keywords on every test run and finally assert the title on the Google search results page. TestNG is the unit-testing framework used here to prioritize and execute tests. The following script is built based on the simple Data-Driven approach using JXL API:

```java
public class SimpleDataDriven {
  WebDriver driver;
  Sheet s;

  @BeforeTest
  public void setUp() {
    driver = new ChromeDriver();
    driver.get("https://www.google.com/");
  }

  @Test(priority = 1)
  public void Googlepass() throws Exception {
    FileInputStream fi = new FileInputStream("/Users/.../data.xls");
    Workbook w = Workbook.getWorkbook(fi);
    s = w.getSheet(0);
    for (int row = 1; row <= s.getRows() - 1; row++) {
```

```
    String input1 = s.getCell(0, row).getContents();
    String output1 = s.getCell(1, row).getContents();

    driver.findElement(By.name("q")).clear();
    driver.findElement(By.name("q")).sendKeys(input1);
    driver.findElement(By.name("q")).sendKeys(Keys.ENTER);
    Thread.sleep(3000);
    try {
      Assert.assertEquals(output1, driver.getTitle());
    } catch (Error e) {
      verificationErrors.append(e.toString());
    }
  }
}


@Test(priority = 2)
public void Googlefail() throws Exception {
  FileInputStream fi = new FileInputStream("/Users/.../data.xls");
  Workbook w = Workbook.getWorkbook(fi);
  s = w.getSheet(0);
  for (int row = 1; row <= s.getRows() - 1; row++) {
    String input2 = s.getCell(2, row).getContents();
    String output2 = s.getCell(3, row).getContents();

    driver.findElement(By.name("q")).clear();
    driver.findElement(By.name("q")).sendKeys(input2);
    driver.findElement(By.name("q")).sendKeys(Keys.ENTER);
    Thread.sleep(3000);
    try {
      Assert.assertEquals(output2, driver.getTitle());
    } catch (Error e) {
      verificationErrors.append(e.toString());
    }
  }
}
}
```

Data-Driven testing using reusable library

The JExcel library does way more than the simple Data-Driven methodologies
explained in the above section. It lets you perform any Java-oriented Data-Driven
tasks by creating a reusable library file. The reusable library is in the form of a class
rather than a JAR file; it can be easily understood and customized based upon the
given specifications:

Create a library file with the following guidelines:

1. Locate and initialize the Excel Binary workbook.
2. Obtain the worksheet and read the Excel sheet row count.
3. Create a function to read the cell value.
4. Create a dictionary using the Hash table and store the Excel sheet column names.

The following is the reusable library class file (`ExcelSheetDriver.java`) built using JXL API:

```java
import jxl.Sheet;
import jxl.Workbook;
import jxl.read.biff.BiffException;

import java.io.File;
import java.io.IOException;
import java.util.Hashtable;

public class ExcelSheetDriver {

    // create sheet name
    static Sheet wrksheet;
    // create workbook name
    static Workbook wrkbook = null;
    static Hashtable dict = new Hashtable();
    //Create a Constructor
    public ExcelSheetDriver(String ExcelSheetPath) throws BiffException,
IOException {
        //Initialize the workbook
        wrkbook = Workbook.getWorkbook(new File(ExcelSheetPath));

        //Here, the worksheet is pointed to Sheet1
        wrksheet = wrkbook.getSheet("Sheet1");
    }

    //Returns the Number of Rows
    public static int RowCount() {
        return wrksheet.getRows();
    }

    //Returns the Cell value by taking row and Column values as argument
    public static String ReadCell(int column, int row) {
        return wrksheet.getCell(column, row).getContents();
```

```
}

//Create Column Dictionary to hold all the Column Names
public static void ColumnDictionary() {
  //Iterate through all the columns in the Excel sheet and store
  the  value in Hashtable
  for (int col = 0; col < wrksheet.getColumns(); col++) {
    dict.put(ReadCell(col, 0), col);
  }
}

//Read Column Names
public static int GetCell(String colName) {
  try {
    int value;
    value = ((Integer) dict.get(colName)).intValue();
    return value;
  } catch (NullPointerException e) {
    return (0);
  }
}
}
```

Now, let's create a test class to perform a simple Google search by reading values from the Excel sheet. To do so, first create a constructor to initialize the Excel data source and then create a loop to iterate through the Excel sheet cell values. The following screenshot is an Excel data source of the preceding example:

	A	B
1	keyword1	keyword2
2	prashanth sams	selenium essentials
3	seleniumworks	bypasshacker
4		

Here, we do a simple Google search for a set of keywords from the Excel file. TestNG is the unit-testing framework used here for executing tests. The following script (Googlesearch.java) is built based on the Data-Driven reusable library using JXL API.

```
public class GoogleSearch {

  //Global initialization of Variables
  static ExcelSheetDriver xlsUtil;
```

```
    //Constructor to initialze Excel for Data source
    public GoogleSearch() throws BiffException, IOException {
      xlsUtil = new ExcelSheetDriver("/Users/.../data.xls");
      //Load the Excel Sheet Col in to Dictionary
      xlsUtil.ColumnDictionary();
    }


    private WebDriver driver;


    @BeforeTest
    public void setUp() throws Exception {
      driver = new ChromeDriver();
      driver.get("https://www.google.com/");
    }


    @Test
    public void Test01() throws Exception {

      //Create a for loop to iterate through the Excel sheet
      for (int rowCnt = 1; rowCnt < xlsUtil.RowCount(); rowCnt++) {

        driver.findElement(By.name("q")).clear();

        //Enter search keyword by reading data from Excel
        driver.findElement(By.name("q")).sendKeys(xlsUtil.
ReadCell(xlsUtil.GetCell("keyword1"), rowCnt));
        driver.findElement(By.name("q")).sendKeys(Keys.ENTER);
        Thread.sleep(2000);

        driver.findElement(By.name("q")).clear();

        //Enter search keyword by reading data from
        Exceldriver.findElement(By.name("q")).sendKeys(xlsUtil.
ReadCell(xlsUtil.GetCell("keyword2"), rowCnt));
        driver.findElement(By.name("q")).sendKeys(Keys.ENTER);
        Thread.sleep(2000);
      }
    }
}
```

Data-Driven testing using TestNG with the @dataProvider annotation

TestNG is a Java-based advanced unit-testing framework similar to JUnit 4, as mentioned in our previous chapters. In this method, we use the TestNG @dataProvider annotation to fetch keywords from an Excel sheet and pass arguments into the test method. This framework using DataProvider is highly recommended for use and an advised Data-Driven framework for implementing Java-based Selenium tests. Let's discuss this with a couple of methods in this section.

In the first method (with Excel), the Excel Binary workbook is used as a data source. This method is usually applied on projects with high data volumes. Here, the TestNG annotation, @dataProvider, allows parameters to pass through the whole iteration process on the Test method. The following screenshot is an Excel data source of the given example:

	A	B	C	D	E	F
1	Test01	column1	column2	column3	column4	
2		selenium Essentials	prashanth sams	selenium works	bypasshacker	
3		one	two	three	four	
4						Test01
5						

Create a test class by getting the array of tables in a two-dimensional array as shown in the following code:

```
public class DataDrivenWithExcel {
  WebDriver driver;
  private String baseUrl;

  @DataProvider(name = "Test")
  public Object[][] createPayId() throws Exception {
    Object[][] retObjArr = getTableArray("C:\\...\\data.xls",
"Sheet1", "Test01");
    return (retObjArr);
  }

  @BeforeClass
  public void BeforeClass() {
    driver = new FirefoxDriver();
    driver.manage().window().maximize();
    baseUrl = "http://www.google.co.in";
  }
```

```
    @Test(dataProvider = "Test", description = "Testing ")
    public void Test01(String column1, String column2, String column3,
String column4) throws Exception {

        driver.get(baseUrl + "/");
        driver.findElement(By.name("q")).sendKeys(column1);
        driver.findElement(By.name("q")).sendKeys(Keys.RETURN);
        Thread.sleep(2000);
        driver.findElement(By.name("q")).clear();
    }

    public String[][] getTableArray(String xlFilePath, String sheetName,
String tableName) throws Exception {
        String[][] tabArray = null;

        Workbook workbook = Workbook.getWorkbook(new File(xlFilePath));
        Sheet sheet = workbook.getSheet(sheetName);
        int startRow, startCol, endRow, endCol, ci, cj;
        Cell tableStart = sheet.findCell(tableName);
        startRow = tableStart.getRow();
        startCol = tableStart.getColumn();

        Cell tableEnd = sheet.findCell(tableName, startCol + 1, startRow +
1, 100, 64000, false);

        endRow = tableEnd.getRow();
        endCol = tableEnd.getColumn();
        System.out.println("startRow=" + startRow + ", endRow=" + endRow +
", " + "startCol=" + startCol + ", endCol=" + endCol);
        tabArray = new String[endRow - startRow - 1][endCol - startCol -
1];
        ci = 0;

        for (int i = startRow + 1; i < endRow; i++, ci++) {
          cj = 0;
          for (int j = startCol + 1; j < endCol; j++, cj++) {
            tabArray[ci][cj] = sheet.getCell(j, i).getContents();
          }
        }
        return (tabArray);
    }
}
```

In the second method (without Excel), the Test class makes use of a built-in data source within a class. Here, the TestNG annotation @dataprovider plays a major role by maintaining effective Data-Driven tests as said in the preceding section. This method is applicable only to projects with lightweight data. Create a Test class by storing data in memory with a two-dimensional array, as shown in the following code:

```java
public class DataDrivenWithoutExcel {
  WebDriver driver;
  private String baseUrl;

  @DataProvider(name = "Test")
  public String[][] y() {

    return new String[][] {
      {
        "prashanth sams", "prashanth"
      }, {
        "selenium esentials", "selenium"
      }, {
        "seleniumworks", "prashanth sams"
      }
    };
  }

  @BeforeClass
  public void BeforeClass() {
    driver = new FirefoxDriver();
    driver.manage().window().maximize();
    baseUrl = "http://www.google.co.in";

  }

  @Test(dataProvider = "Test")
  public void Google(String actual, String expected) throws Exception
  {

    driver.get(baseUrl + "/");
    driver.findElement(By.name("q")).sendKeys(actual);
    driver.findElement(By.name("q")).sendKeys(Keys.RETURN);
    Thread.sleep(2000);
    boolean b = driver.getPageSource().contains(expected);
    Assert.assertTrue(b);
```

```
    }

    @AfterClass
    public void AfterClass() {
      driver.quit();
    }

}
```

Apache POI Data-Driven framework

Apache **Poor Obfuscation Implementation** (**POI**) is a set of Java-based libraries used to manipulate Microsoft Excel documents, such as the .xls and .xlsx file formats. It is an extensive API used to automate Selenium Data-Driven tests that lets you create, modify, read, and write Excel data. Unlike JXL API, it supports both Binary and SpreadsheetML workbooks. Configuring JXL API is easier compared to Apache POI. However, Apache POI has as many features to work with modern Microsoft products. Obviously, the test performance with the .xlsx file will be slower compared to the .xls file in Apache POI.

A dependency of Apache POI is the xmlbeans library, which has to be added in the build path before executing tests. Another Java framework for older versions of Apache POI is dom4j. The following screenshot displays a list of library files required to support the Apache POI Data-Driven framework:

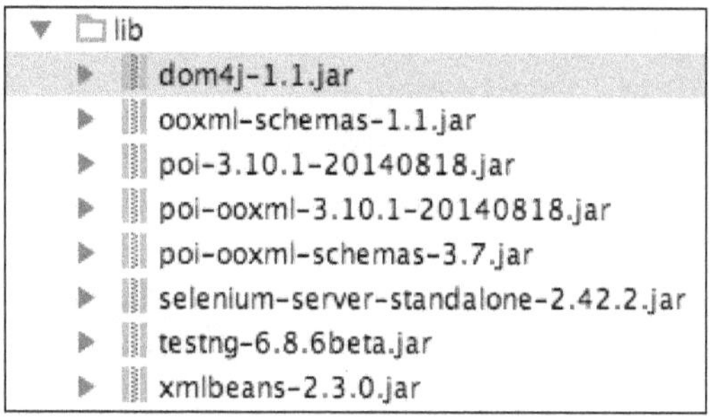

The preceding JAR files are the significant libraries to be added in the build path. Here, the workbook is a common interface for the HSSF, XSSF, and SS usermodels. Refer to the following snippets to write values in different workbooks:

```
Workbook wb = new HSSFWorkbook(); // HSSF Workbook
FileOutputStream fo = new FileOutputStream("/Users/.../workbook.xls");
wb.write(fo);
fo.close();
```

```
Workbook wb = new XSSFWorkbook(); // XSSF Workbook
FileOutputStream fo = new FileOutputStream("/Users/.../workbook.
xlsx");
wb.write(fo);
fo.close();
```

Apache POI provides significant features such as creating a new workbook, new sheet, new cells, and date cells; formatting cells; setting footers; setting zoom; freezing panes; and enhancing cell styles using colors, fonts, and borders.

The following code snippet helps to create a new worksheet:

```
InputStream fi= new FileInputStream("/Users/.../workbook.xls");
// SS Usermodel
Workbook wb = WorkbookFactory.create(fi);
Sheet ws= wb.getSheetAt(0);

// HSSF Usermodel
Workbook wb = new HSSFWorkbook();
Sheet ws = wb.createSheet("Sheet1");

// XSSF Usermodel
Workbook wb = new XSSFWorkbook();
Sheet ws = wb.createSheet("Sheet1");
```

HSSF usermodel – Binary workbook

Horrible SpreadSheet Format (HSSF) implements the Excel 97 Binary workbook with the `.xls` file format. Apache POI has specific features similar to JXL API that read and write Excel workbooks at any instance of time. HSSF takes cell data in the cell format and converts format, such as numeric, string, Boolean, or formula, into string, as shown in the following code:

```
HSSFCell cell = row.getCell(j);
String value = cellToString(cell);
```

The two-dimensional array retrieves the external data source by storing all values from an Excel sheet into the data array. Here, `data` is the variable and `i` and `j` are the row and column numbers:

```
String[][] data = new String[rowNum][colNum];
data[i][j] = value;
```

Let's discuss with an example how to read and write values in the Excel Binary workbook (.xls). The following example is customized based on the TestNG unit testing framework with the @dataProvider annotation to iterate through the list of values available in a spreadsheet. The following screenshot is an Excel data source of the given example.

	A	B
1	keyword1	keyword2
2	selenium essentials	selenium essentials
3	selenium works	prashanth sams

Build a test class using the hssf.usermodel.* library classes from the Apache POI API. The following script depicts a simple Google search by reading a set of keywords and writing the PASS / FAIL status in the same Binary worksheet:

```java
import org.apache.poi.hssf.usermodel.HSSFCell;
import org.apache.poi.hssf.usermodel.HSSFRow;
import org.apache.poi.hssf.usermodel.HSSFSheet;
import org.apache.poi.hssf.usermodel.HSSFWorkbook;
import org.apache.poi.ss.usermodel.Cell;
import org.apache.poi.ss.usermodel.Row;


public class Excelreadwrite {

  private static WebDriver driver;
  private static String baseUrl;
  private static int n = 0;

  @BeforeTest
  public void setUp() {
    driver = new ChromeDriver();
  }

  @Test(dataProvider = "DP")
  public static String login(String keyword1, String keyword2)
  throws Exception {

    driver.get("https://www.google.com");

    driver.findElement(By.name("q")).sendKeys(keyword1);
    driver.findElement(By.name("q")).sendKeys(Keys.ENTER);
    Thread.sleep(2000);
    String actualtitle = driver.getTitle();
    System.out.println(actualtitle);
```

```
    driver.findElement(By.name("q")).clear();

    driver.findElement(By.name("q")).sendKeys(keyword2);
    driver.findElement(By.name("q")).sendKeys(Keys.ENTER);
    Thread.sleep(2000);
    String expectedtitle = driver.getTitle();
    System.out.println("Expected Title is: " + expectedtitle);
    driver.findElement(By.name("q")).clear();

    int LastRow = ++n;
    if (expectedtitle.equals(actualtitle)) {
      System.out.println("PASSED");
      String status = "PASS";
      excelwrite(status, LastRow);
    } else {
      System.out.println("FAILED");
      String status = "FAIL";
      excelwrite(status, LastRow);
    }
    return expectedtitle;
  }

public static void main(String[] args) throws Exception {
    excelRead();
  }

@DataProvider(name = "DP")
public static String[][] excelRead() throws Exception {
    File excel = new File("/Users/.../data.xls");
    FileInputStream fis = new FileInputStream(excel);
    HSSFWorkbook wb = new HSSFWorkbook(fis);
    HSSFSheet ws = wb.getSheet("Sheet1");
    int rowNum = ws.getLastRowNum() + 1;
    int colNum = ws.getRow(0).getLastCellNum();
    String[][] data = new String[(rowNum - 1)][colNum];
    int k = 0;
    for (int i = 1; i < rowNum; i++) {
      HSSFRow row = ws.getRow(i);
      for (int j = 0; j < colNum; j++) {
        HSSFCell cell = row.getCell(j);
        String value = cellToString(cell);
        data[k][j] = value;
      }
      k++;
```

```
      }
      return data;
    }


    public static void excelwrite(String status, int LastRow) throws
Exception {
        try {
          FileInputStream file = new FileInputStream(new File("/Users/.../
data.xls"));

          HSSFWorkbook workbook = new HSSFWorkbook(file);
          HSSFSheet sheet = workbook.getSheetAt(0);

          Row row = sheet.getRow(LastRow);

          Cell cell2 = row.createCell(2); // Shift the cell value
depending upon column size
          cell2.setCellValue(status);
          // System.out.println(status);
          file.close();
          FileOutputStream outFile = new FileOutputStream(new File("/
Users/.../data.xls"));
          workbook.write(outFile);

        } catch (FileNotFoundException e) {
          e.printStackTrace();
        } catch (IOException e) {
          e.printStackTrace();
        } catch (HeadlessException e) {
          e.printStackTrace();
        }
    }


    public static String cellToString(HSSFCell cell) {
      int type;
      Object result;
      type = cell.getCellType();
      switch (type) {
        case 0:
          result = cell.getNumericCellValue();
          break;
        case 1:
          result = cell.getStringCellValue();
```

```
        break;
      default:
        throw new RuntimeException("There are no support for this type
of cell");
      }
    return result.toString();
  }
}
```

Let's create the `testNG.xml` file inside the project folder to execute the preceding test cases. Here, the `.xml` file allows you to define the package name, class, groups, methods, include, exclude, and much more. Invoking `testNG.xml` is an efficient method of running TestNG tests. See the following example to run the test:

```
<?xml version="1.0" encoding="UTF-8" ?>
<!DOCTYPE suite SYSTEM "http://testng.org/testng-1.0.dtd">
<suite name="Suite" allow-return-values="true" parallel="none">
  <test name="Test">
    <classes>
      <class name="packagename.Excelreadwrite" />
    </classes>
  </test>
  <!-- Test -->
</suite>
<!-- Suite -->
```

The following screenshot is a report obtained after running the preceding test case. You will notice a new column with values inserted (PASS or FAIL) based on the status after test completion, as follows:

	A	B	C
1	keyword1	keyword2	
2	selenium essentials	selenium essentials	PASS
3	selenium works	prashanth sams	FAIL

For more information on HSSF Read Binary (`.xls`) workbook, refer to `http://goo.gl/YqfzvZ`
The `TestNG.xml` file for the preceding example is available at `http://goo.gl/0KyaSR`.

XSSF usermodel – SpreadsheetML workbook (.xlsx)

The **XSSF** (**XML SpreadSheet Format**) usermodel implements Excel 2007, a SpreadsheetML workbook with the OOXML (.xlsx) file format. Handling tests through the HSSF API on Excel 2003 results in running out of memory on the creation of large sheets. Apache POI introduces the XSSF API to avoid such risks by handling Excel 2007. Let's see an example of reading values from the Excel SpreadsheetML workbook. This example is customized based on the TestNG unit-testing framework using @dataProvider annotation to iterate through the list of values available in a spreadsheet. The following screenshot is an Excel data source of the given example:

	A	B
1	keyword1	keyword2
2	selenium essentials	seleniumworks
3	prashanth sams	bypasshacker

In this example, we perform a simple Google search by reading a set of keywords from the SpreadsheetML workbook through the Apache POI API:

```java
import org.apache.poi.xssf.usermodel.XSSFCell;
import org.apache.poi.xssf.usermodel.XSSFRow;
import org.apache.poi.xssf.usermodel.XSSFSheet;
import org.apache.poi.xssf.usermodel.XSSFWorkbook;

public class className {
    private WebDriver driver;
    private String baseUrl;
    private StringBuffer verificationErrors = new StringBuffer();

    @BeforeTest
    public void setUp() throws Exception {
        baseUrl = "http://www.google.com/";
        driver = new FirefoxDriver();
        driver.manage().window().maximize();
        driver.manage().timeouts().implicitlyWait(30, TimeUnit.SECONDS);
    }

    @Test
    public void Test01() throws Exception {
        driver.get(baseUrl + "/");
        InputStream inp = new FileInputStream("/Users/.../data.xlsx");
```

```
XSSFWorkbook wb = new XSSFWorkbook(inp);
XSSFSheet ws = wb.getSheet("Sheet1");

for (int i = 1; i <= ws.getLastRowNum(); i++) {
  XSSFRow row = ws.getRow(i);
  XSSFCell col1 = row.getCell(0);
  XSSFCell col2 = row.getCell(1);

  driver.findElement(By.name("q")).sendKeys("" +col1+ "");
  driver.findElement(By.name("q")).sendKeys(Keys.ENTER);
  Thread.sleep(2000);
  driver.findElement(By.name("q")).clear();
  driver.findElement(By.name("q")).sendKeys("" +col2+ "");
  driver.findElement(By.name("q")).sendKeys(Keys.ENTER);
  Thread.sleep(2000);
  driver.findElement(By.name("q")).clear();
    }
  }
}
```

SS usermodel – Binary and SpreadsheetML workbooks

The SS usermodel (`org.apache.poi.ss.usermodel`) is common for both the Binary (`.xls`) and SpreadsheetML (`.xlsx`) workbooks; however, the HSSF and XSSF usermodels focus on a single workbook. So there we have it:

```
SS = HSSF + XSSF
```

The SS usermodel is the most commonly used API and the most easy-to-handle workbook. Unlike HSSF and XSSF, the SS usermodel uses the `Factory` class, **org. apache.poi.ss.usermodel.WorkbookFactory**, to handle workbooks. The following is the syntax for this usermodel:

```
Workbook wb = WorkbookFactory.create(filepath);
```

Let's discuss with an example to read values from either Excel Binary or SpreadsheetML workbooks. The following screenshot is an Excel data source of the given example:

	A	B
1	keyword1	keyword2
2	selenium essentials	seleniumworks
3	prashanth sams	bypasshacker

Build a test class using `ss.usermodel.*` library classes from the Apache POI API. The following script depicts a simple Google search by reading a set of keywords from any kind of Excel worksheet:

```
import org.apache.poi.ss.usermodel.*;

public class className {
  private WebDriver driver;
  private String baseUrl;
  private StringBuffer verificationErrors = new StringBuffer();

  @BeforeTest
  public void setUp() throws Exception {
    baseUrl = "http://www.google.com/";
    driver = new FirefoxDriver();
    driver.manage().window().maximize();
    driver.manage().timeouts().implicitlyWait(30, TimeUnit.SECONDS);
  }

  @Test
  public void Test02() throws Exception {
    driver.get(baseUrl + "/");
    InputStream fp = new FileInputStream("/Users/.../data.xlsx");

    Workbook wb = WorkbookFactory.create(fp);
    Sheet ws = wb.getSheetAt(0);

    Row row = null;

    for (int i = 1; i <= ws.getLastRowNum(); i++) {
      row = ws.getRow(i);
      Cell col1 = row.getCell(0);
      Cell col2 = row.getCell(1);

      driver.findElement(By.name("q")).sendKeys("" + col1 + "");
      driver.findElement(By.name("q")).sendKeys(Keys.ENTER);
      Thread.sleep(2000);
      driver.findElement(By.name("q")).clear();
      driver.findElement(By.name("q")).sendKeys("" + col2 + "");
      driver.findElement(By.name("q")).sendKeys(Keys.ENTER);
      Thread.sleep(2000);
      driver.findElement(By.name("q")).clear();

      row = null;
```

```
            }
        }
    }
```

Text file Data-Driven framework

The text file is a human-readable plain-text file with the `.txt` file extension. Unlike Microsoft documents, the text file doesn't need any external APIs to support Data-Driven tests. However, it depends on the individual to develop their own framework to handle a text file.

To separate two or more values in a text file, detach the keywords on customizing suitable expressions, such as comma, space, and so on, as follows:

```
String[] data = line.split(", ");
```

Notepad is a default application normally used to access the text file. Certainly, the text file is a universal file, which is platform independent and can be accessed from any machine. A minimum version of Java 7 is mandatory to execute tests based on the text file. Let's discuss with an example how to drive tests using the text file as a data source. The following screenshot is a text data source of the given example:

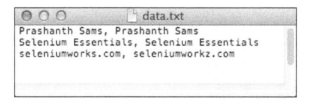

This example covers a Google search accompanied by an assertion between the expected value and the actual value on the Google results page. Let's see the following complete workaround:

```
import java.io.BufferedReader;
import java.io.File;
import java.io.FileInputStream;
import java.io.InputStreamReader;

public class className {
    WebDriver driver;
    private String baseUrl;

    @BeforeClass
    public void BeforeClass() {
```

```
    driver = new FirefoxDriver();
    driver.manage().window().maximize();
    baseUrl = "http://www.google.co.in";
}


@Test
public void Test01() {
  Execute();
}


public void Execute() {

  BufferedReader in = null;
  InputStreamReader inputStream = null;
  try {

      inputStream = new InputStreamReader(new FileInputStream("/
Users/.../" + File.separator + "data.txt")); in = new
BufferedReader(inputStream);
      String line = null;
      String actualvalue = "";
      String expectedvalue = "";
      while ((line = in .readLine()) != null) {
        String[] data = line.split(", ");
        if (data.length >= 1) {
          actualvalue = data[0];
          expectedvalue = data[1];
          System.out.println("Actual : " + actualvalue);
          System.out.println("Expected : " + expectedvalue);
          driver.get(baseUrl + "/");
          WebElement element = driver.findElement(By.name("q"));
          element.sendKeys(actualvalue);
          element.sendKeys(Keys.RETURN);
          boolean b = driver.getPageSource().contains(expectedvalue);
          Assert.assertTrue(b);
        }
      }
    } catch (Exception ex) {
      ex.printStackTrace();
    }
  }
}
```

Data-Driven testing using TestNG with the @dataProvider annotation – text file

Annotation is one of the fascinating features introduced by the TestNG framework that was later implemented in JUnit 4. In fact, the annotations from both the frameworks always differ. In this method, we are using the `@dataProvider` annotation of TestNG to fetch multiple sets of data from a text file and pass arguments to a test method. Meanwhile, it's pretty safe to place the text file inside your project. The following syntax helps you to do it:

```
System.getProperty("user.dir")
```

Define a constant header to tag the row values. Here, search1 and search2 are the two constants used. The following screenshot is a text data source of the given example:

The following example is a simple Google search that reads a set of keywords separated by a comma and a space using reusable `TextDriver`. Create a test class (`GoogleTest.java`) as follows:

```java
import java.util.HashMap;

public class GoogleTest {
  private WebDriver driver;

  @DataProvider(name = "keywords")
  public Object[][] data() throws Exception {
    HashMap < String, String[] > dataSet = new TextDriver(System.
getProperty("user.dir") + "/data.txt").getData();

    String search1Strings[] = dataSet.get("search1");
    String search2Strings[] = dataSet.get("search2");
    int size = search1Strings.length;

    // modify 2 upon the no. of rows; Here, I used two rows,
    'search1'&'search2'
    Object[][] creds = new Object[size][2];
    for (int i = 0; i < size; i++) {
```

```
        creds[i][0] = search1Strings[i];
        creds[i][1] = search2Strings[i];
      }
      return creds;
    }


    @BeforeTest
    public void setUp() throws Exception {
      driver = new ChromeDriver();
    }


    @Test(dataProvider = "keywords", description = "Google_Test")
    public void search(String search1, String search2) throws Exception
    {

      driver.get("http://www.google.co.in");

      // search google with keyword1
      driver.findElement(By.name("q")).clear();
      driver.findElement(By.name("q")).sendKeys("" + search1);
      driver.findElement(By.name("q")).submit();
      Thread.sleep(4000);

      // search google with keyword2
      driver.findElement(By.name("q")).clear();
      driver.findElement(By.name("q")).sendKeys("" + search2);
      driver.findElement(By.name("q")).submit();
      Thread.sleep(4000);
    }
  }
```

As discussed previously in the *JXL API* section, the reusable library used here is a customized class but for a text file that can be modified and used instead of an external library. Here, the one-dimensional array stores value and splits them accordingly, as follows:

```
String[] keyValue = stringLine.split(" = ");
keyValuePair.put(keyValue[0], keyValue[1].split(", "));
```

The following is a reusable library class file (TextDriver.java) to fetch data from a text file:

```
import java.io.BufferedReader;
import java.io.FileInputStream;
import java.io.FileReader;
```

```
import java.util.HashMap;

public class TextDriver {

  private String fileLocation;

  public TextDriver(String fileLocation) {
    this.fileLocation = fileLocation;
  }

  public HashMap < String, String[] > getData() {
    FileInputStream fs;
    HashMap < String, String[] > keyValuePair = new HashMap < String,
String[] > ();
    try (BufferedReader br = new BufferedReader(new
FileReader(fileLocation))) {
      String stringLine;
      //Read File Line By Line
      while ((stringLine = br.readLine()) != null) {
        // Print the content on the console
          String[] keyValue = stringLine.split(" = ");
          keyValuePair.put(keyValue[0], keyValue[1].split(", "));
      }

    } catch (Exception e) {
      e.printStackTrace();
    }
    return keyValuePair;
  }
}
```

Properties file Data-Driven framework

The properties file is similar to the text file but with the .properties file extension. It is widely used in Java apps to store and configure parameters.

Data-Driven testing using the properties file as the data source is feasible on Selenium WebDriver to handle small data. Create a .properties file and store the keywords with a constant key (search). In Eclipse, open the project, right-click on the src folder and select **New** | **Other...** | **General** | **File**. Name the file with the . properties extension, for example, config.properties.

Let's see an example of how to read values from the properties file:

```
config.properties          ×
1   search = seleniumWorks, prashanthSams, seleniumEssentials
2
```

Build a test class that depicts a simple Google search by reading a series of keywords from the `.properties` file:

```java
import java.util.ResourceBundle;
import java.util.StringTokenizer;
import java.util.concurrent.TimeUnit;

public class classname {
  private WebDriver driver;
  private String baseUrl;

  @BeforeTest
  public void setUp() throws Exception {
    driver = new ChromeDriver();
    baseUrl = "https://www.google.co.in";
  }

  @Test
  public void Test01() throws Exception {

    ResourceBundle bundle = ResourceBundle.getBundle("config");
    String Channel = bundle.getString("search");
    StringTokenizer st = new StringTokenizer(Channel, ", ");
    while (st.hasMoreTokens()) {
      String value = st.nextToken();
      driver.get(baseUrl + "/");
      driver.findElement(By.name("q")).click();
      driver.findElement(By.name("q")).sendKeys(value);
      driver.findElement(By.name("q")).sendKeys(Keys.ENTER);
      Thread.sleep(2000);
    }
  }
}
```

Let's see an alternative method to fetch values from the `.properties` file. More than a data source, the properties file acts as a repository by storing web elements for data reusability. The following is the syntax for the properties file:

```
prop.getProperty("path")
```

The following screenshot is a properties data source of the next example. Here, q is a path as mentioned in the preceding syntax and search is a constant key to define the search keyword.

```
config.properties        ×
1    search = seleniumEssentials
2    path = q
```

The following example contains a simple Google search functionality driven with the Java utility library on a properties file:

```java
import java.io.FileInputStream;
import java.util.Properties;
import java.util.concurrent.TimeUnit;

public class className {
  private WebDriver driver;
  private String baseUrl;

  @BeforeTest
  public void setUp() throws Exception {
    driver = new ChromeDriver();
    baseUrl = "https://www.google.co.in";
  }

  @Test
  public void Test01() throws Exception {
    driver.get(baseUrl + "/");

      FileInputStream fs = new FileInputStream("/Users/.../config.
properties");
      Properties prop = new Properties();
      prop.load(fs);

      String value = prop.getProperty("search");
      System.out.println(prop.getProperty("search"));
      driver.findElement(By.name(prop.getProperty("path"))).click();
      driver.findElement(By.name(prop.getProperty("path"))).
sendKeys(value);
      driver.findElement(By.name(prop.getProperty("path"))).
sendKeys(Keys.ENTER);
      Thread.sleep(2000);
    }
}
```

Data-Driven testing using TestNG with @dataProvider annotation – properties file

In this method, we use the `@dataProvider` annotation of the TestNG framework to fetch multiple sets of data from a properties file and pass arguments to a test method.

Create a `.properties` data source file similar to the following screenshot, where the data is separated with a comma and a space (`, `). The two constant keys, `search1` and `search2`, are used to define the data available in a `properties` file.

```
data.properties                    ×
1   search1=Prashanth Sams, Selenium Essentials
2   search2=seleniumworks.com, bypasshacker
```

The following example is a simple Google search that reads a set of keywords separated by a comma and a space using `PropertiesDriver` that is reusable. Create a test class (`GoogleTest.java`), as follows:

```java
import java.util.HashMap;

public class GoogleTest{
  private WebDriver driver;

  @DataProvider(name = "keywords")
  public Object[][] data() throws Exception {
  HashMap<String, String[]> dataSet = new PropertiesDriver().
getData();

    String search1Strings[] = dataSet.get("search1");
    String search2Strings[] = dataSet.get("search2");
    int size = search1Strings.length;

    // modify 2 upon the no. of rows; Here, two rows, 'search1'&
    'search2' are used
    Object[][] creds = new Object[size][2];
    for (int i = 0; i < size; i++) {
      creds[i][0] = search1Strings[i];
      creds[i][1] = search2Strings[i];
    }
    return creds;
  }

  @BeforeTest
  public void setUp() throws Exception {
    driver = new ChromeDriver();
```

```
    }

  @Test(dataProvider = "keywords", description = "Google_Test")
  public void search(String search1, String search2) throws Exception
{

    driver.get("http://www.google.co.in");

    // search google via keyword 1
    driver.findElement(By.name("q")).clear();
    driver.findElement(By.name("q")).sendKeys("" + search1);
    driver.findElement(By.name("q")).submit();
    Thread.sleep(2000);

    // search google via keyword 2
    driver.findElement(By.name("q")).clear();
    driver.findElement(By.name("q")).sendKeys("" + search2);
    driver.findElement(By.name("q")).submit();
    Thread.sleep(2000);
  }
}
```

The following is a reusable library class file (PropertiesDriver.java) to fetch data from a properties file:

```
import java.util.Enumeration;
import java.util.HashMap;
import java.util.ResourceBundle;

public class PropertiesDriver {
  public PropertiesDriver() {
  }
  public HashMap<String, String[]> getData() {
  HashMap<String, String[]> configMap = new HashMap<String,
String[]>();
  try {
    ResourceBundle bundle = ResourceBundle.getBundle("config");
      Enumeration<String> keys = bundle.getKeys();
      while (keys.hasMoreElements()) {
        String aKey = keys.nextElement();
        String aValue = bundle.getString(aKey);
        configMap.put(aKey, aValue.split(","));
      }
    } catch (Exception e) {
      e.printStackTrace();
```

```
    }
    return configMap;
  }
}
```

CSV file Data-Driven framework

The term CSV refers to **Comma-Separated Values**. In CSV, plain-text values are automatically accessed as tabular data. The following screenshot depicts the CSV data source. These values are actually stored using a notepad with separator (comma) between any two keywords.

	A	B	C
1	Selenium Essentials	Prashanth Sams	seleniumworks
2			

The `FileReader` class is a Java library class used to read data from a CSV file. Refer to the following code snippet that explains how to read data from a CSV-formatted file.

```
String path = "/Users/.../data.csv";
File file = new File(path);
BufferedReader IN = new BufferedReader(new FileReader(file));
String line = null;
while ((line = IN.readLine()) != null) {
  String[] data = line.split(",");
  driver.findElement(By.locatorType("path")).sendKeys(data[0]);
  driver.findElement(By.locatorType("path")).sendKeys(data[1]);
}
```

The `FileWriter` class is a Java library class to write data into a CSV file. This library class helps you to create a CSV file and store output data as a fresh copy. The following code snippet explains how to write data into a CSV file:

```
FileWriter writer = new FileWriter("/Users/.../output.csv");
writer.append("your_text or runtime_value");
```

Let's see an example that involves both reading and writing CSV files. The following example depicts a simple Google search by reading a set of keywords and writing the attained page title into a CSV file.

```
import java.io.BufferedReader;
import java.io.File;
import java.io.FileReader;
import java.io.FileWriter;
```

```java
public class classname {
  private WebDriver driver;
  private String baseUrl;

  String path = "/Users/.../data.csv";

  @Before
  public void setUp() throws Exception {
    driver = new ChromeDriver();
    baseUrl = "https://www.google.com";
  }

  @Test
  public void Test01() throws Exception {
    driver.get(baseUrl + "/");

    FileWriter writer = new FileWriter("/Users/.../output.csv");
    writer.append("ColumnHeader1");
    writer.append(',');
    writer.append("ColumnHeader2");
    writer.append('\n');

    File file = new File(path);
    BufferedReader IN = new BufferedReader(new FileReader(file));
    String line = null;
    while ((line = IN.readLine()) != null) {
      String[] data = line.split(",");

      driver.findElement(By.name("q")).sendKeys(data[0]);
      driver.findElement(By.name("q")).submit();
      Thread.sleep(2000);
      String element1 = driver.getTitle();
      driver.findElement(By.name("q")).clear();

      driver.findElement(By.name("q")).sendKeys(data[1]);
      driver.findElement(By.name("q")).submit();
      Thread.sleep(2000);
      String element2 = driver.getTitle();

      writer.append(element1);
      writer.append(',');
      writer.append(element2);
      writer.append('\n');
      writer.flush();
```

```
      }
      try {
        IN.close();
      } catch (Exception e) {
        System.out.println(e);
      }
    }
  }
```

Keyword-Driven framework

A Keyword-Driven test is a kind of functional automation testing, where keywords are used instead of scripts. Since all the required Selenium functions and operations are prewritten in an external user-defined driver, basic knowledge of the framework's workflow is more than enough to learn and maintain tests. Furthermore, a fully developed Keyword-Driven framework reduces a tester's scripting effort. Let's take a look at Open2Test, which is purely a Keyword-Driven framework that supports Selenium WebDriver too. This framework handles the Binary Excel sheet through JExcel API.

The Open2Test components, namely `Selenium_Utility`, `ObjectRepository`, `TestSuite`, and `Test_Script` are the source files used to perform Keyword-Driven tests. See the following screenshot to get an idea about the framework structure:

These components are available in the form of Excel sheets, where Utility Excel stores all the location paths of the object repository, test suite, test script, and report folder. The following table represents Selenium Utility Excel of the Open2Test framework:

File \ Folder Name	Path Location
Test Suite	`C:/.../Test_Suite.xls`
Test Script	`C:/.../Test_Script.xls`
Object Repository	`C:/.../Object_Repository.xls`
Summary Report	`C:/.../Report/`
Screen shot Report	`C:/.../Screen_shot/`
Detailed Report	`C:/.../Detailed_Report/`

Let's see how to use this approach in an example with a simple Google search. Create an object repository Excel sheet, where the elements can be stored with an object name. These objects are always reusable and maintained in a separate Excel file. The following is an object repository Excel content table that contains elements of a Google page:

Object name	Object type	Parent	Object path
`clear_search`	Textbox	1	`name=q`
`google_search`	Textbox	1	`name=q`
`submit_search`	Button	1	`id=sblsbb`

As we know, a test suite is a collection of test cases; it can have any number of test cases. The following table is a test-suite Excel content table with a couple of test scripts:

Run	Test scripts
r	`Test_Script.xls`
r	`Test_Script2.xls`

The test case sheet contains keywords to perform Selenium WebDriver functions, such as click, clear, submit, and so on. Here, the keywords, such as `launchapp`, `perform`, `check`, `condition`, `storevalue`, and `loop` are predefined in the Open2Test framework that give you full control over the application under test. The following test case (`Test_Script.xls`) Excel content table contains a clean Google search workflow:

Run	Keyword	Object details	Action
r	`launchapp`	`http://www.google.com`	
r	`Perform`	`Textbox;clear_search`	`Clear`
r	`Perform`	`Textbox;google_search`	`set:Selenium Essentials`
r	`Perform`	`Button;submit_search`	
r	`Wait`	`5000`	

An Open2Test keyword is **launchapp**, which allows users to access the Google page URL. The `Perform` keyword is used to execute Selenium WebDriver functions. Meanwhile, there are keywords that assert or verify an element. Remember that this framework has limitations and is not yet fully stable. Certainly, the Open2Test framework can be modified with additional functions.

 To acquire the Open2Test driver, refer to `https://bit.ly/1yz5FOL`.

Creating a Keyword-Driven framework consumes more time and scripting skills; however, it is highly efficient and reliable. Moreover, it reduces the scripting effort and improves the reusability of test scripts.

Hybrid-Driven framework

A combination of the Data-Driven and Keyword-Driven (or Modular-Driven) frameworks is commonly said to be a Hybrid-Driven framework. In general, a Hybrid-Driven framework is a collection of two or more frameworks that can be customized and accessed by any user. For example, a combination of PageObjects, a Keyword-Driven framework, a Data-Driven framework, an object repository, and reporting listeners provides a powerful Hybrid framework.

Let's take a tour of the framework that can be helpful as a part while building a Hybrid approach.

Create a driver class (`loadDriver.java`) that loads user-defined methods by invoking Selenium WebDriver functions:

```
public class loadDriver {

  private static WebDriver driver;

  public static void Firefox() {
    driver = new FirefoxDriver();
    driver.manage().window().maximize();
    System.out.println("Firefox browser is initiated...");
  }
  public static void IE() {
    driver = new InternetExplorerDriver();
    driver.manage().window().maximize();
    System.out.println("Internet Explorer is instantiated...");
  }
  public static void Chrome() {
```

```
        driver = new ChromeDriver();
        driver.manage().window().maximize();
        System.out.println("Google Chrome is instantiated...");
    }
    public static void URL() {
        driver.get(ObjectRepository.URL);
    }
    public static void waitForID(String id) {
        WebDriverWait wait = new WebDriverWait(driver, 10);
        wait.until(ExpectedConditions.presenceOfElementLocated(By.
id(id)));
    }
    public static void WaitForPageLoad() {
        driver.manage().timeouts().pageLoadTimeout(10, TimeUnit.SECONDS);
    }
    public static void clearID(String id) {
        driver.findElement(By.id(id)).clear();
    }
    public static void submit(String id) {
        driver.findElement(By.id(id)).submit();
    }
    public static void assertPageTitle() {
        Assert.assertEquals(driver.getTitle(), ObjectRepository.title);
    }
    public static void insertText(String id, String searchValue) {
        driver.findElement(By.id(id)).sendKeys(searchValue);
    }
    public static void exit() {
        driver.close();
        driver.quit();
    }
    public static void sleeper(int value) throws InterruptedException {
        Thread.sleep(value);
    }
}
```

In the preceding code, the driver automatically shrinks the code size in a test class and maintains an object repository to store entire web elements for data reusability. These practices help you to optimize reusability and make compact test scripts. Shown here is an object repository for Google search:

```
public class ObjectRepository {
    public static final String URL = "https://www.google.com/";
    public static final String searchField = "lst-ib";
    public static final String searchText1 = "Prashanth Sams";
```

```
      public static final String searchText2 = "Selenium Essentials";
      public static final String submitButton = "sblsbb";
      public static final String title = "Prashanth Sams - Google Search";
}
```

Create a test class (GoogleTest.java) with a couple of use cases. It involves both negative and positive workarounds. TestNG is a framework used here to prioritize and execute test methods. The following is the code for the test class with the use cases:

```
public class GoogleTest {

  //Positive Use_case
  @Test(enabled = true, priority = 2)
  public void TC_01() throws InterruptedException {
    loadDriver.Chrome();
    loadDriver.URL();
    loadDriver.WaitForPageLoad();
    loadDriver.waitForID(ObjectRepository.searchField);
    loadDriver.clearID(ObjectRepository.searchField);
    loadDriver.insertText(ObjectRepository.searchField,
    ObjectRepository.searchText1);
    loadDriver.submit(ObjectRepository.submitButton);
    loadDriver.sleeper(3000);
    loadDriver.assertPageTitle();
    loadDriver.exit();
  }

  //Negative Use_case
  @Test (enabled = true, priority = 1)
  public void TC_02() throws InterruptedException {
    loadDriver.Chrome();
    loadDriver.URL();
    loadDriver.WaitForPageLoad();
    loadDriver.waitForID(ObjectRepository.searchField);
    loadDriver.clearID(ObjectRepository.searchField);
    loadDriver.insertText(ObjectRepository.searchField,
    ObjectRepository.searchText2);
    loadDriver.submit(ObjectRepository.submitButton);
    loadDriver.sleeper(3000);
    loadDriver.assertPageTitle();
    loadDriver.exit();
  }
}
```

Summary

In this chapter, we learned about the different types of Selenium frameworks with examples, how to build automation frameworks from scratch, and how to optimize a successful Selenium WebDriver automation framework.

Index

HTMLUnit 36, 39
HTMLUnitDriver 39
Hybrid-Driven framework 166-168

I

IEDriverServer
about 26
IE, instantiating with 26
URL 27
iframes
handling 106, 107
isElementPresent method 82, 83

J

Java Excel API (JXL API) 133
Java Robot
URL, for keyboard actions 112
used, for handling browser pop-up 108
used, for handling native OS 108
JavascriptExecutor class
about 114
elements, highlighting 117
JavaScript error collector 118, 119
new browser window, opening 118
page scroll methods 115
JavaScript functions, Selenium IDE
about 15
for mouse scroll 16
parameterization, with arrays 17, 18
runScript command 16
JBehave
about 128
embedders 128
injectable embedders 128
URL 129
JBehave BDD framework
about 128
using 129-132
JSErrorCollector library
about 118, 119
URL, for downloading 119
JUnit-enabled embeddables
JUnitStory 129
JUnitStories 129

JXL API Data-Driven framework
about 133
Data-Driven testing, reusable library
used 137-139
Data-Driven testing, using TestNG
with @dataProvider 141-143
data, reading in Excel sheet 133, 134
data, writing in Excel sheet 133, 134
simple Data-Driven approach 136

K

keyboard actions 73
Keyword-Driven framework
about 164-166
framework structure 164

L

launchapp 166
Linux
ChromeDriver, configuring 7
locator prioritization 8, 9

M

Mac
ChromeDriver, configuring 8
Maven 123
Mofiki's Coordinate Finder 112
mouse and keyboard actions
accept() method 72
authenticateUsing() method 72
build() method 73
click() method 73
clickAndHold() method 74
contextClick() method 74
dismiss() method 71
doubleClick() method 74
dragAndDrop() method 75
dragAndDropBy() method 75
getText() method 72
keyDown() method 75
keyUp() method 76
moveByOffset() method 76
moveToElement() method 76

Thank you for buying
Selenium Essentials

About Packt Publishing

Packt, pronounced 'packed', published its first book, *Mastering phpMyAdmin for Effective MySQL Management*, in April 2004, and subsequently continued to specialize in publishing highly focused books on specific technologies and solutions.

Our books and publications share the experiences of your fellow IT professionals in adapting and customizing today's systems, applications, and frameworks. Our solution-based books give you the knowledge and power to customize the software and technologies you're using to get the job done. Packt books are more specific and less general than the IT books you have seen in the past. Our unique business model allows us to bring you more focused information, giving you more of what you need to know, and less of what you don't.

Packt is a modern yet unique publishing company that focuses on producing quality, cutting-edge books for communities of developers, administrators, and newbies alike. For more information, please visit our website at www.packtpub.com.

About Packt Open Source

In 2010, Packt launched two new brands, Packt Open Source and Packt Enterprise, in order to continue its focus on specialization. This book is part of the Packt Open Source brand, home to books published on software built around open source licenses, and offering information to anybody from advanced developers to budding web designers. The Open Source brand also runs Packt's Open Source Royalty Scheme, by which Packt gives a royalty to each open source project about whose software a book is sold.

Writing for Packt

We welcome all inquiries from people who are interested in authoring. Book proposals should be sent to author@packtpub.com. If your book idea is still at an early stage and you would like to discuss it first before writing a formal book proposal, then please contact us; one of our commissioning editors will get in touch with you.

We're not just looking for published authors; if you have strong technical skills but no writing experience, our experienced editors can help you develop a writing career, or simply get some additional reward for your expertise.

Selenium WebDriver Practical Guide

ISBN: 978-1-78216-885-0   Paperback: 264 pages

Interactively automate web applications using Selenium WebDriver

1. Covers basic to advanced concepts of WebDriver.

2. Learn how to design a more effective automation framework.

3. Explores all of the APIs within WebDriver.

4. Acquire an in-depth understanding of each concept through practical code examples.

Selenium Design Patterns and Best Practices

ISBN: 978-1-78398-270-7   Paperback: 270 pages

Build a powerful, stable, and automated test suite using Selenium WebDriver

1. Keep up with the changing pace of your web application by creating an agile test suite.

2. Save time and money by making your Selenium tests 99% reliable.

3. Improve the stability of your test suite and your programing skills by following a step-by-step continuous improvement tutorial.

Please check **www.PacktPub.com** for information on our titles

Selenium Testing Tools Cookbook

ISBN: 978-1-84951-574-0   Paperback: 326 pages

Over 90 recipes to build, maintain, and improve test automation with Selenium WebDriver

1. Learn to leverage the power of Selenium WebDriver with simple examples that illustrate real world problems and their workarounds.

2. Each sample demonstrates key concepts allowing you to advance your knowledge of Selenium WebDriver in a practical and incremental way.

3. Explains testing of mobile web applications with Selenium Drivers for platforms such as iOS and Android.

Selenium 2 Testing Tools Beginner's Guide

ISBN: 978-1-84951-830-7   Paperback: 232 pages

Learn to use Selenium testing tools from scratch

1. Automate web browsers with Selenium WebDriver to test web applications.

2. Set up Java Environment for using Selenium WebDriver.

3. Learn good design patterns for testing web applications.

Please check **www.PacktPub.com** for information on our titles

www.ingramcontent.com/pod-product-compliance
Lightning Source LLC
Chambersburg PA
CBHW060130060326
40690CB00018B/3819